UNTETHERED

UN TETHERED

HOW VISIONARIES THINK
ABOUT THE MOBILE-FIRST FUTURE

SHAWN BORSKY

FOREWORD BY JUSTIN NGUYEN

UNTETHERED
How Visionaries Think about the Mobile-First Future

FIRST EDITION

ISBN   978-1-5445-4727-5   *Hardcover*
       978-1-5445-4726-8   *Paperback*
       978-1-5445-4775-6   *Ebook*

To La and Toby,

For all the great childhood summers in Florida.

Badass role model and power woman before that was a thing, La is also a wonderful human—kind, calm, and thoughtful. If I make it into my nineties like she has and am half the person she is, I'll die happy.

Not by blood but by heart, Toby was my bombastic, kind, and friendly "Grandpa." His empathy, humility, no-nonsense business acumen, and one-liners still echo in my mind every day. And yes, I've thought, thought again, checked, and then done.

CONTENTS

FOREWORD 9

INTRODUCTION.. 13

1. IT'S PERSONAL.. 19

2. IT'S NOT ABOUT YOU 33

3. MEET THEM WHERE THEY ARE............................ 43

4. MEET THEM AS THEY ARE 55

5. HAVE SOMETHING THEY WANT 65

6. HELP THEM MAKE IT THEIRS 75

7. INTEGRATE WITH THEIR LIVES 85

8. DELIGHTFUL!... 93

CONCLUSION 103

ACKNOWLEDGMENTS................................. 107

FOREWORD

—JUSTIN NGUYEN, DECEMBER 2022

I met Shawn for the first time in 2015 when he walked into my office at Blizzard Entertainment for a job interview. I'd come to Blizzard from Electronic Arts the year before and had just taken charge of its modest mobile division as part of a recent company reorg. There were larger roles available on our web teams, but I had asked for mobile. I believed it was poised to deliver on all the shiny futuristic promises made in the '90s by the advent of the internet. At last, we were going to get the information, tools, services, and experiences we wanted whenever we wanted them, wherever we were. I wanted to push Blizzard onto mobile's frontlines. All I needed was a designer who saw what I did to partner with me to make it happen.

In walked Shawn. Gregarious and excitable, he answered the question, "What does mobile mean to you?" with a breadth of vision and depth of knowledge that blew me away.

When he left, the hiring manager looked like a man who'd surfed a tornado. "One to five," he asked me, "What do you think?"

"Six," I said. "Whatever it takes, we have to get that guy on our team."

And we did. Over the next five years, Shawn and I built an organization inside Blizzard that innovated, managed, and delivered eight unique mobile applications across multiple platforms.

In the years since, Shawn's reputation as the mobile designer other designers turn to has grown significantly. Not even a decade later, the demand for his consulting expertise on what he aptly calls the "untethered internet" is such that he's had to reach back to the older technology of books to meet it. I'm glad he has. It's comforting to me that Shawn's design philosophy is in such high demand. As a mobile user, I know if I start interacting with any of the companies he's advised, I'm going to have a well-designed, untethered experience.

Today, mobile isn't the future anymore. It hasn't even just arrived. It's dominant. Mobile devices are in the hands of more people than any other Information Age technology. The amount of time we spend on smartphones dwarfs that spent on computers, and the disparity will only increase. The reason is as unchanging as its expression is dynamic: every business in the history of business has had the same goal—to get more people doing more of what it wants them to do more frequently while spending more time and money doing it. Nothing delivers that better than the untethered internet. But only when appropriately deployed.

No matter how expensive it is, if a company's mobile presence or a UX designer's interface includes dropdown pick lists or hamburger menus (that useless but ubiquitous little three-line icon), their mobile engagement is going to be lower than it could be. They're doing the equivalent of using this flashlight as a candle holder because they can't be bothered with batteries.

Untethered is your box of AAs. As Shawn's eight attributes of mobile-native design explain, there are things a website just can't do. Unlike mobile website design which ignores them, native mobile development puts the attributes of the untethered internet that have made it so popular with consumers to work for businesses.

This is why if you're a tech developer, product manager, product owner, designer, or software engineer, your professional future is untethered. Likewise, a CEO, CFO, CMO, CPO, CXO, VP, director, or general manager who doesn't understand untethered technology is the constraint on their business's ability to create new value. Success in the mobile world is our generation's golden egg-laying goose. All it takes is an entirely new mindset. It's a bit of a chasm, but if you haven't crossed it yet, you're restricting your professional growth or company's potential. Probably both. I hope you'll put Shawn's batteries in your flashlight and see those eggs shine.

INTRODUCTION

In 2007, BlackBerry was sitting at the top of the personal handheld device market, cheerily unconcerned with things like music and photos. You know the story; it regularly gets trotted out as a cautionary tale about the dangers of corporate hubris. BlackBerry's leaders got complacent about its feature set. They weren't looking for ways to evolve or exploring how their customers interacted with their devices. BlackBerry's leaders thought they understood their business. They made pagers and had advanced that technology as far as it could reasonably go. Sure, there would be tweaks and upgrades, but there was no credible risk of disruption by a superior pager.

In a way, they were right. But they were still in trouble.

BlackBerry was threatened—not by technological stagnation or even hubris but by a very human difficulty: the inability to imagine a near future that isn't a marginally improved-upon version of the present. Missing the implications of a new technology on an existing, seemingly unrelated one isn't a modern problem. Our ancestors probably didn't expect fire's domestication to impact much beyond how they ate their mammoth steaks. Media theorist Marshall McLuhan (of "the medium is the message" fame) dubbed it the Horseless Carriage Syndrome and used it to explain why early movies look like filmed stage plays and why early websites were little more than online brochures. BlackBerry's leaders knew about Apple, but they weren't worried about a computer company's entry into the handheld market. They couldn't conceptualize a device that wasn't a pager challenging their market share. The iPod was for music. It didn't even have a keyboard. Not a threat.

Fourteen years after the iPhone toppled BlackBerry, 85 percent of Americans have a smartphone, and many business leaders can't yet see how that has anything much to do with them. Sure, they know "reaching customers online" no longer means exclusively through a website viewed on a desktop computer. They may even agree that the future is mobile. But like BlackBerry back in 2007, most companies are carrying the present's mental models into their vision of tomorrow. They think of their mobile presence as a miniaturized version of their website (with maybe a branded app thrown in for good measure). They think "mobile" refers only to smartphones.

I get it. The change has happened quickly and seems like a linear progression. When the internet stopped being a bulletin board on the roadside of the "information superhighway" and first caught the attention of forward-thinking companies, it didn't pose a huge imaginative leap. It created more interactive, closer to real-time versions of existing business assets. Websites and email were easy to understand as electronic mail and digital product catalogs. Customers still connected to a company from their desks, and a company reached its customers through mass marketing in the traditional media of print and broadcast TV or radio.

The connection started to thin with the widespread switch from desktop to laptop computers and from phone lines to Wi-Fi. The quick adoption of cell phones and then smartphones stretched it even further. But the future of mobile isn't just a longer umbilicus—it's untethered. People are no longer connected; they're *within*. Like from pager to smartphone, the difference isn't one of degree.

To the digital native, "mobile" doesn't mean phone. It's not a dictionary, library, music catalog, or anything else "in your pocket." It's not about communication, information, or accessing information (and everything else!) "on the go." Untethered users expect their experience and information to move fluidly with them through their smartphone, their smartwatch, or even their car, adapting itself to these and multiple other devices and to their varying levels of attention. Accordingly, companies are no longer entities to them but environments within the information flow that they inhabit. As such, they want relationships, not transactions.

This is great news. Untethered consumers are more loyal and less likely to churn. They're also the future, and business leaders who want to be a part of it need a new mental model. They need to reconceptualize "mobile" as an atmosphere within which their brand has multiple opportunities to enhance its relationship with its customers by being where it's needed in whatever capacity people want, regardless of their location, platform, or device and sensitive to the fluctuating time, attention, and memory users have available in those different contexts.

The companies that are successfully designing for this untethered future today are more focused on their users' experience than their company's products and on delivering value rather than capturing eyeballs. The absolute best of them are also deeply integrated into people's lives. They're platform- and device-agnostic, and they provide multiple points of contact. They make it easy for users to customize their experiences and understand their customers' needs, wants, and desires, and they adapt to users' changing levels of memory and attention. Ideally, they also make those interactions delightful.

The untethered, mobile-first future is as different from the early, desktop-bound internet as a Tesla is from a Model T. In *Untethered*, I'll reframe what mobile means, explore what it makes possible, and introduce a new way of thinking about it. More philosophy than a how-to, *Untethered* will propose a new way of navigating the untethered future by filtering it through eight key attributes.

UNTETHERED DESIGN IS:

1. Non-transactional
2. Empathetic
3. Application-, device- and platform-agnostic
4. Flexible
5. Relevant
6. Personalized and customizable
7. Integrated
8. (and for bonus points) Delightful

I'll devote a chapter to each attribute and explain what it is, how it works, and how it interacts with the other attributes to create the untethered experience. Many of the companies I'll hold up as examples are, unsurprisingly, tech companies (Tesla, Apple, Facebook, and Netflix, among others), but there are also standouts in industries as diverse as food service, sports equipment, and garage door openers. *Untethered* offers a way of conceptualizing mobile technology that leaders in any industry can apply and entrepreneurs can use to shape their new businesses.

Untethered won't necessarily teach you to be forward-thinking, but it will (I hope) change how you think about new products and invest in mobile. It also will give you a set of criteria to test your existing products against. This list isn't prescriptive, and a product doesn't need to have all eight attributes to be untethered, but many of today's most successful offerings meet at least two-thirds of the list. Of course, not every product or service can be untethered, but thinking about your business with an untethered mindset will help you keep up with the technology and consumers' changing expectations and demands.

Having worked in mobile and online gaming since its earliest days, I've had a front-row seat for the advent of the untethered future. I've also contributed to it—from the early days working to get higher-quality 3D games onto flip phones through doing development for dedicated iPhone games. I helped create ways of interacting with Blizzard games via app and text message and contributed to several award-winning mobile-app-based companies that had no other significant online presence. As my understanding of what untethered technology makes possible expanded, and as I distilled it into a design philosophy, I found myself in demand as a teacher and consultant.

In that work, I was fortunate to work with many visionary leaders who "got it" right away, but I also had a lot of frustrating experiences. I've seen decision-makers unable to shift from thinking by linear analogy (i.e., "mobile is like digital, only cordless") to reconceptualizing it. It's my hope that having it dissected and explained here will help make it more accessible and keep all but the leading edge from falling behind.

Because cars weren't an incremental change from horses, early car companies couldn't imagine a freeway. Too many leaders today are missing opportunities on the edges of what an untethered internet makes possible because they're limited by the capabilities of the previous technology. In the same way that the horseless carriage created car culture, mobile technology is giving rise to its own ethos. The untethered philosophy of mobile design isn't just theoretical. It's a prediction of a future that, as sci-fi writer William Gibson pointed out, "is already here—it's just not evenly distributed yet." Interestingly, for all its technological advancement, perhaps mobile design's most defining feature is how human-centric it is. The core principle of a successful business is still "give the people what they want," and what the people want is personal.

IT'S PERSONAL

"Can I borrow your phone?"

Even if there's nothing embarrassing on it, you'll probably hesitate. If we know each other well, you might feel okay with letting me queue up a song or key in my number. But you'll probably want it back quickly, and you wouldn't want me clicking out of whatever app you opened before handing it over. There's nothing unusual about your reluctance. Most people feel the same way.

It is, however, a strange (or at least a new) way to feel about a piece of technology. I doubt you have any qualms about who watches your TV. Pretty much anyone you trust enough to allow into your house is likely to be left alone in the den with it if they need to check on the game or amuse their children. Yet once upon a time, we arranged our music on shelves spines out for visitors to scan, and we borrowed phones at bars or hotels if there wasn't a public payphone nearby. Yesterday's phones felt more public. Today's are private.

People tend to have a more personal, one-to-one relationship with untethered devices. When I tap the email icon on my smartphone, it opens *my* email. Spotify brings up my play-lists. Even my banking app requires very little proof that I'm authorized to check my balances. Multifactor identification is predicated on the assumption that only I have my phone. I am myself on my smartphone—a person—and I expect to be treated that way.

I'm not alone. Because people have a more personal, one-to-one connection with untethered devices than with other technology, consumers expect the companies they interact with inside that space to treat them like people and to behave more like people themselves.

BEING SOCIAL

Mark Zuckerberg may have started Facebook to help Harvard students connect with each other, and Twitter defines itself as "home to a diverse world of people," but almost every company has a Facebook page it wants you to like. Most companies tweet. Many respond directly to individual customer comments on both platforms. Some are hysterically inept, and many more come across as tone-deaf or irrelevant. Almost universally, the businesses that have been successful on Twitter, Facebook, and other social media channels have been the ones that use them to interact with customers as individuals and that behave more like individuals themselves. Social media isn't inherently social or personal. Truly personal interactions (the kind untethered users expect) need to be more—a series of mutually beneficial exchanges built on trust, over time, in which not every interaction is a transaction.

Untethered consumers want relationships with companies that are mutually beneficial, based on shared values, and developed over time.

The untethered consumer expects to have strong human-like relationships with the companies they support. The strongest marker of such relationships (and most long-term relationships, really) is that they are non-transactional.

BE NON-TRANSACTIONAL

I'm a big fan of Nike's run-tracking app. Not only does it monitor where, how far, and how efficiently I run, but it also helps me connect with other runners and provides me with a great example of the first untethered principle. It is a well-funded mobile-first app that offers tremendous value to users for free. It is beautifully non-transactional.

The difference between a transactional and a relational company is a bit like the difference between eating to live and living to eat. A transactional company seeks to minimize the time and expense required to generate a sale. A relational company invests in repeat customers. It's the reason your local bar might comp you the last drink of the night while a hotel lobby bar likely will not. One is hoping you'll come back and thus invests in the relationship. The other isn't and doesn't. Likewise, if every time my wife asked me to get something for her while I was up, I responded with, "Sure! What will you do for me in return?" my marriage would sour quickly. Human beings seem to keep some sort of internal score, but it's impressionistic more than tallied. It's unclear when an acquaintance becomes a friend or when a passing fancy becomes love, but you know it when you see it.

UNTETHERED IS INFINITE

In his fascinating book, *The Infinite Game*, Simon Sinek makes a distinction between games played to win (finite ones) and those you play to keep playing (infinite games). Finite games have clearly defined success criteria and a clear beginning and end. They have winners and losers. Football, like most formal sports, is a finite game. Infinite games may have rules, but they don't have winners or losers. "The floor is hot lava" is an infinite game. If you touch the carpet, you scream in fake pain, but you jump back on the sofa and keep going.

Sinek proposes that most successful businesses today are playing an infinite game. Rather than positioning themselves to beat the competition, they're focused on keeping their company, their culture, and their value "in the game." As contrasting examples, he holds up Apple and Microsoft. Having attended the conferences of both companies, he found Apple focused on the future and Microsoft focused on Apple. While both companies have their own diehard fans, Sinek argues that Apple's fan base is broader and its culture happier and more sustainable because it's playing an infinite game.

If consumers were strictly rational, nobody would buy an Apple laptop. But they do. In part, I believe, because they perceive Microsoft as far more transactional. From how lovingly its products are packaged to the quasi-exclusivity of its app store, Apple *feels like* it cares about your experience. It cultivates a feeling in its users that it wants to go beyond whatever they might have expected. It creates the expectation that if users remain loyal and keep investing in its ecosystem, they will continue to have their expectations exceeded and their lives enhanced.

Today's untethered consumers don't mind paying for what they want—nobody expects Nike to give away sneakers—but they do expect some investment in them in return. They expect reciprocity.

By providing consumers with something of value for free, Nike creates something of an emotional open loop in much the same way that you might be inspired to drop a donut by the desk of the colleague who brought you cookies last week. (Before you both go out for a run to burn off the extra calories, of course.)

Nike wants consumers to associate its brand with their individual and personal fitness journey. If it can add value to something that matters to you, your perception of Nike will be more positive. The Nike brand will also more readily come to mind when you think about fitness-related purchases. But of course, fostering a non-transactional relationship with customers isn't required for success in the untethered landscape. You don't have to look any further than Facebook or Amazon for proof.

People regularly talk *on Facebook* about how much they hate Facebook, and I know several people who apologize for shopping on Amazon. But they still shop there. All but the most committed will continue to use companies they suspect may be exploiting them if the transactional value is great enough or if the company has an effective monopoly. If Facebook is the only place you can connect with Aunt Marge, get updates from your favorite band, or sign up for the next yoga class, you're going to use Facebook. But if another company comes along that delivers the same value, I believe the people who feel used by Facebook or railroaded by Amazon will leave in droves.

Nike, on the other hand, has done quite well despite the many comparable products available. Because Nike has invested in the relationship it has with people, they're more loyal to it. There is no other way to build loyalty. The only thing you have that another company can never deliver to your customers is the relationship you have with them.

Non-transactional relationships build loyalty.

Companies that have relationships with their customers have an advantage, even over competitors with equal or higher-quality products and equal or lower prices because people put a value or premium on that relationship. Relationships tend to mean shared values, and seeing something of yourself in a company always drives more value in that connection.

USP OR POV?

Traditionally, marketing has been tasked with informing consumers about what a product is able to deliver. It highlights the product benefit or unique selling proposition (USP)—the thing only this product can deliver. In purely utilitarian transactions, this makes excellent sense. If all a customer cares about is price and you have the lowest prices, advertising that fact is good for business. Point of view (POV) marketing, however, doesn't focus on the distinguishing features of an offering but on the company's ideology about it. Apple doesn't advertise its specs or price; it advertises its philosophy of what technology should be.

This doesn't mean that every company needs to have a point of view on everything, but it does require consistency. USPs can be values-neutral, but relationships rarely are. People who love a company can come to hate it. A point of view that's just marketing can do more damage than having none at all. A company whose hiring practices discriminate against LGBTQIA+ job seekers should not try to sell Pride-branded T-shirts.

If you promote your company by claiming to have certain values, your customers will expect you to act accordingly and react very negatively when you do not. People don't want to be treated like a number, but we'd rather be treated like numbers than idiots.

In both transactions and relationships, people expect honesty—or at least hold out a skeptical hope for it. If you sell me a dozen donuts and there are only eleven in the box, I'll be angry. If you sell me "artisanal, organic doughnuts that contribute to the local economy" and you're secretly a multinational donut conglomerate, the betrayal is the same.

BRAND EQUITY

You can measure the power of a brand by the number of missteps it can absorb. A first date who's twenty minutes late may not get a second date, but a spouse of ten years is unlikely to go home alone for the same infraction. If my favorite local restaurant serves me something subpar, I'm likely to give it another chance. If it makes me violently ill, no amount of loyalty is going to get me back in the door (although I have returned to a cafe I like after a light scuffle with food poisoning).

Nike had enough brand equity to recover from the sweatshop scandal, but had it not also made significant changes to the way it sourced and produced its products, I doubt it would have survived. The untethered consumer has a finely calibrated BS detector, and authenticity is a dangerous thing to fake, but people aren't naïve about what keeps the cost of shoes and clothing low. The issue wasn't really that Nike was using sweatshops. While that was certainly immoral, the outrage came from how misaligned using them was with Nike's brand—a brand built on shared values.

SHARED VALUES

Nike has long been a frontrunner (forgive me) in recognizing the benefits of fostering a relationship with its customers around a set of shared values. Nike publicly supports inclusive charities and gets its brand on inspirational athletes, knowing these are ideals their target customers support and want to be associated with. If the Nike brand represents inclusive and inspiring physical performance, people with similar aspirations, wanting to associate themselves with the things they believe in, are likely to support the brand. Some of this is reciprocity (*Nike supports the things I value, so I support it*), and some of it is more self-directed (*I buy Nike because it makes me feel good about myself*).

> People prefer to see something of themselves in the companies they support, and they like doing business with companies that make them feel good about themselves.

Of course, companies have blurred the line between charitable giving and paid marketing since the "brought to you by Ovaltine" days. Law firms fund the local opera to support the arts and burnish their image. Anheuser-Busch sponsors NASCAR for at least similar reasons. Red Bull pours money into extreme sports, and Pepsi spends billions on pop star endorsements for brand exposure, but neither sponsors NPR. It isn't just about visibility; it's about alignment.

Nike not only states support of its mission; it demonstrates that support by spending enormous resources on the charities, community programs, and athletes that embody it. The company has even developed and built an entire website dedicated to such advocacy. While not pure altruism, it does demonstrate a commitment to a mission that its customers are also committed to. It's effective because Nike's commitment is demonstrated rather than announced. The company puts its money where its mouth is, investing time and treasure in things that don't have an obvious return on investment but add to the perception that Nike isn't just "in it for the money." It's in it to "bring inspiration and innovation to every athlete in the world." (Nike's lovely annotation to "athlete," which both strengthens and broadens its brand, says, "if you have a body, you're an athlete.")

Many companies understand the value of building brand goodwill. This isn't a new idea, but few have extended that thinking to untethered technology. This doesn't mean every company that wants to reach the untethered consumer needs to invest in developing an app (more on this in Chapter 3). Companies need to adopt the same mindset—thinking about how they can add value to their customers' lives and what might create and foster relationships between them. Luckily, the technology that makes this necessary also makes it easy. Or at least easier.

Mobile has made it much easier and cheaper to have multiple small interactions rather than single transactions. In fact, the value untethered technology is best equipped to deliver isn't sales but engagement, but most companies aren't set up to build relationships. Their very organizational structure inhibits it. Without a visionary leader to create incentives for making an organization more humanlike, people within it act like people and look out for their own interests. To build the kind of humanlike companies that untethered consumers increasingly expect, leaders need to rewrite their success criteria and reward engagement.

REWRITE SUCCESS CRITERIA

Obviously, making money is part of any company's definition of success, and the shift from transactional to relational isn't one away from commercial success. It's more long-tail commercial. Investing in the sponsorship of charities that promote inclusion and athletes whose stories feature the single-minded focus to "just do it" every day for years is an *investment*, not a sacrifice.

Too many e-commerce sites define success by how quickly they convert visits to sales. In other words, the more quickly a person makes a purchase, the more successful the company believes its website to be. So what's the success metric for Nike's run-tracking app? People interact with it regularly, spend a long time with it, and never make a purchase. By first-generation e-commerce definitions, it's a failure. Nike created the app in the (correct) belief that it would contribute to a brand that would attract sales—but by working toward that goal indirectly through relationship-building.

Traditional e-commerce success: How do I get customers to convert to immediately meet my business goals?

Untethered mobile-first success: How do I deliver value to my customers to eventually meet my business goals?

REWARD ENGAGEMENT

Most companies split product development into design and engineering, with the design team focused on *what* to create and the engineering team working on *how* to create it. Design, often in tandem with a product team, looks at things like problem sets, product features, and market requirements. It's their job to synthesize the customer's needs, wants, and desires (more on this in the next chapter) with the company's business goals.

In this model, a product manager whose performance is judged on the number of sneakers sold isn't likely to suggest plowing resources into a run-tracking app that doesn't sell sneakers. If mobile isn't a great place to sell sneakers, they're not terribly interested in it. There's no incentive for them to be. To encourage people within an organization to do the things that create companies that act like people, you need an incentive structure that rewards increases in engagement as well as sales.

While it's easier than ever to know how often an online interaction results in a sale, the origin of that sale is harder to determine. But it's not impossible. Nike can (and I'm sure does) evaluate the number of purchases made by people who use its app against those who don't. But there's no doubt that investing in mobile and increased engagement requires longer-range and larger thinking because the increases it drives in sales and customer loyalty aren't obvious in immediate, one-to-one correlations. Still, Nike's run-tracking app may be free, but I'll bet it pays for itself.

WHERE THE CUSTOMER JOURNEY BEGINS

Nike probably still tracks conversions on its website, but it recognizes that arriving at Nike.com is not the start of most customer journeys. Likewise, an untethered e-commerce company might still measure success by how quickly a visitor to its site makes a purchase, but that's not the sole criterion for success. The company is interested in whether the customer responds to educational material it may have sent via email or to an advertising campaign tied to Pride month. The company builds out sections of its website that aren't focused on driving purchases but on providing useful information or encouraging people to interact in some other way. Even if the company's business model is based on website product sales, it's optimized to work better on a mobile device and to deliver content and add value before trying to get you to buy something. It's relational, not transactional.

SUMMARY

Perhaps the foundational principle of untethered mobile is a shift in the relationship between company and consumer from purely transactional to mutual investment. It's business, but it's personal. While not every company needs to have a human relationship with its consumers, it's the only way to create customer loyalty and build up the kind of brand equity that provides resilience against missteps and setbacks. Social media is an obvious arena for companies to act more like people, but a clever Twitter feed won't do it any more than knowing a lot of jokes will guarantee a solid friendship.

There are two primary ways a company can be more like a person: by engaging in non-transactional interactions with customers and by demonstrating shared values. Happily, untethered technology is the ideal mechanism for both, but leveraging that capacity requires a new understanding of the customer journey and rewriting success criteria.

Untethered customers want to have a relationship with the companies they support, but they won't just give you their money because they like you. You have to deliver something of value.

CHAPTER 2

IT'S NOT ABOUT YOU

When I go out for a run today, I won't strap on a pedometer and heart rate monitor, thread my headphones through my clothes, or find that one shirt with a special pocket for my smartphone. I leave my phone at the house because a profile on my headphones "knows" I'm going running, connects to my smartwatch (which holds my music library), cues up an appropriate playlist, and syncs with an app that tracks my route and monitors my stats.

On my run, if I happened to trip and fall into the proverbial ditch, Apple's "hard fall detection" feature would kick in. Reacting to information from my smartwatch's built-in accelerometer, it would first ping me. If I didn't respond, it would quickly contact emergency services in my area with information about my condition and location. It would then notify the people on my emergency contact list and let them know I might be in trouble. Strava could also reach out to nearby runners. It's a feature that's saved several lives and reassured countless adult children of aging but quite active parents (I'm looking at you, Dad).

The hard fall detection hasn't saved my life yet, but it does provide a beautiful example of what is often called "design consciousness," or more commonly "design thinking." This is the should-be-obvious-but-is-more-rare-than-it-should-be practice of putting the user at the center of product design and development. Apple can provide a lifesaving feature people didn't know they wanted because it's invested heavily in developing an empathetic understanding of their needs, wants, and desires.

BE EMPATHETIC

In the last chapter, we talked about the importance of making investments *in* as well as sales *to* customers by delivering a constellation of value. Because knowing what's of value to people requires empathy, untethered companies must be empathetic.

This is often much harder than it sounds. Companies tend to conflate their customers' goals with their business goals and can struggle with distinguishing between what they want, what their customers want, and what they want their customers to want.

BUSINESS VERSUS CONSUMER GOALS

In his wonderful book *Alchemy,* Rory Sutherland describes an awkward meeting with British Airways. The company had recently made a multimillion-dollar investment in new aircraft, and it was his uncomfortable task to convince them they would be better off advertising their cucumber sandwiches than their planes. Most customers don't know the difference between an Airbus Seven-something-something and a Boeing Something Else. They do readily differentiate between a tiny pretzel bag's inauthentic gesture at snacks and a nice in-flight (and very British) teatime treat.

Likewise, website users want to use a website with little struggle. They don't care about your new color scheme, no matter how much you paid an extra-spicy branding agency to pick it for you. Upgrades to its functionality that only make it easier for customers to give you their money won't have much persuasive impact on them either. Taking the time to stop and ask yourself why customers have come to your site and what their goals are may seem painfully obvious, but it's easy to lose sight of. Alan might think "Alan's" is a great name for his company, but "Alan's Autos" will speak more directly to his customers about what matters to them and include alliteration to boot.

Company mission statements are a terrific place to spot this confusion between customer- and company-focused goals. Becoming the number-one distributor of X or the industry leader in Y, achieving exponential growth, or doubling sales doesn't do much for consumers. It is, however, what consumers can do for you if you're able to meet a need, want, or desire of theirs. Most everyone understands and accepts that businesses are in business to make money, but there's a significant difference between being in the business of making money and making money in the business of providing something people need, want, or desire.

Even companies that understand, at the macro level, the value of being customer-focused still struggle to execute accordingly in their day-to-day business operations. Within organizations that track quarterly financial targets and with incentive structures that reward them, most companies make short-term decisions that prioritize revenue. Many others trick themselves by simply framing their business goals in a way that sounds customer-centric while viewing consumers only as means to an end. For example: "Our customers can't be served if they can't buy more of our products" or "Power users will want more premium features that won't feel premium if we don't charge to add them!" *(I really had this conversation with a high-up leader.)*

I'm sure that Apple would like every smartwatch owner to have an Apple laptop, smartphone, and television, but if being everyone's one-stop computing shop was its ultimate aim, it probably wouldn't be the most valuable company in the world. Unfortunately, Apple's success contributes to the confusion. Because Apple is often innovative and being seen as innovative is appealing to businesses and entrepreneurs (particularly in the tech sector), Apple often sets goals like "being the innovation leader in our industry" or "disrupting our sector," not realizing these lofty ambitions are the company's. They're not what consumers need, want, or desire.

INNOVATION

Innovation without empathy is like buying your wife a cutting-edge, propane-fired BBQ grill for her birthday because you want to try cooking with gas. The truth is that most good products aren't particularly innovative because innovation per se isn't something customers need, want, or desire, and people interact only with companies because—you guessed it—they need, want, or desire something.

That said, when innovation is informed by empathy, it can both differentiate a company (at least until the others catch on) and allow it to pivot quickly when circumstances change.

When Starbucks first introduced its in-app ordering for in-store pickup, I was dubious. I could see that it would let them move people into and out of their stores more quickly, but I didn't see what was in it for me. A friend of mine even thought they were being short-sighted. She knew she'd frequently been enticed into a coffee cake she hadn't intended to buy while waiting by the pastry case to place her drink order.

But Starbucks understood us better than we did ourselves and gambled that its caffeine-craving customers tended toward the impatient and liked saving time. After watching mobile order early adopters breeze in, grab their drinks, and leave while we waited in line, the rest of us started to consider our options, and once we tried it, we liked it! It's now more typical for me to order coffee from the quiet comfort of my car, watch a YouTube video on my smartphone, and then stroll in to pick my coffee up than wait in line.

People started putting in huge smartphone orders they would have been too embarrassed or disorganized to put in if they had to do so in person. People experimented with options the app offered for customization and happily saved the best results for easy (and thus more frequent) reordering. It changed some customers' behavior, allowing them to create such insanely complex orders that baristas now have their own subreddit thread to share them. It lets me pick up a coffee in the fifteen minutes I have between meetings when I would not have risked being late by waiting in line. The app even started recommending coffee cake to my friend and reminding me that I often like a breakfast sandwich with my morning black coffee.

Although to-go ordering from restaurants was hardly a new idea, Starbucks was the first large restaurant chain to introduce ordering in-app for pickup in stores. It was an innovation based on an empathetic understanding of its customers. It also allowed the company to pivot quickly in the early days of the COVID-19 pandemic, when people started staying home in droves, by meeting its customers' desire for coffee with their suddenly conflicting need for safety and social distancing. The company streamlined its app, making its food and drinks easier for customers to order and safer for them to pick up. While this was more adaptation than innovation, empathy helped them leverage their existing model to scale into places that weren't their conventionally dominant ones.

As I've been repeating, people interact with companies and products to satisfy a need, want, or desire. Companies focused on meeting one of that power trio often discover new—even innovative—ways of serving their customers. A company focused on innovation itself can almost always find interesting, new things to do, but they're often just solutions looking for problems. Successful companies empathize with their customers and design for them.

EMPATHETIC DESIGN

Most people will tell you that Apple is quite good at design, but the truth and power of that reach well beyond aesthetics. I regularly encounter people who think of design as synonymous with the visual, but Apple doesn't owe its success to its ability to provide simply beautiful or functional hardware. Apple is the most valuable company in the world because of the way its hardware is crafted and integrated into its design. The difference between functional and empathetic design is the difference between holistic and reductive problem-solving. Both ways set out to solve a customer problem, but empathetic design prioritizes solving a problem customers *want* to have solved.

Sometimes, as in the Starbucks example, the design solves a problem the customer knows they have and wants to have solved. But often, empathy allows you to anticipate a problem that customers would want to have solved if they realized such a solution were possible, the way Apple's hard fall detection feature solved the "lying unconscious in a ditch" problem.

Empathetic design allows you to meet non-obvious needs.

Empathy allows companies to meet customer needs not only intentionally but thoughtfully. SpaceX's Dragon capsule has minimal touch screens and carefully rounded corners on its (probably vegan) leather seats; it's a stunning white, futuristic-looking spacecraft. The intention behind the design is methodical and meticulous, which translates to a futuristic, cool-looking aesthetic that generates tons of goodwill. Yes, a less attractive spacecraft could have gotten into orbit (looking at you, NASA space shuttles), but empathetic, intentional design is detail-oriented by nature. The screens and windows are visible proof of the level of care taken with every other element.

The perfectly folded piece of paper around the precisely coiled charging cable that comes with every Apple product is translated by consumers as a symbolic representation of the care they unconsciously assume Apple is taking with everything else. People go back to the companies that have provided them with these experiences because that's how people want to be treated. Intuitively, we want to be cared for by companies that are intentional and thoughtful.

Aesthetic beauty is an effect of (not the reason for) empathetic design. The more empathy, intention, and attention to detail that go into the design of a thing, the more beautiful it will be because we value (and perceive as beautiful) things that result from the expenditure of energy and effort. We know that things that are well-crafted not only perform better, but they take longer to create, and we appreciate the extra effort. Empathetic design understands that appreciation and helps meet a higher tier of user needs.

HIERARCHY OF USER NEEDS

The hierarchy of user needs (or UX Pyramid) closely mirrors Maslow's famous, larger hierarchy of human needs. Products move upward from mere functionality (it works) to reliably functional (it works when I need it to) through usability (I can easily figure out how to make it work and work it easily) to pleasure (It's fun to use). It's a unidirectional pyramid. A friend recently lent me his old pickup to haul some home renovation supplies. It was a beast—no shocks, no radio. No fun at all. The clutch was finicky, and even once I got a feel for it, it wasn't easy to use. I was coasting hard on that *one* driving lesson I had on how to pop a clutch, and I had to leave it running because I'd been warned not to rely on it starting again if I turned it off. It reminded me of my old Chevy Blazer, where I kept a hammer in the glove box so I could hit the starter back into action. But as with Old Blaze, it worked, and its bed could hold a half sheet of drywall. It was functional. A pickup that can't function as a drywall transport device is a de facto failure no matter how pleasurable, usable, *and* reliable it is.

Most companies can deliver functionality and reliability, but usability requires empathetic design. I'm sure the designers of early VHS machines knew exactly how to program their clocks, but they failed to understand the abilities of the average person old enough to afford one. The much more technically complex iPad was immediately usable to my grandmother because of its empathy-enabled intuitive interface. Today, iPads are an increasingly common enrichment device for babies and toddlers—now that's usable!

The top of the hierarchy (and beyond it, from pleasurable to delightful, which we'll save for Chapter 8) is inaccessible without empathy. Arguably, it's the difference between the archetypical old-school PC (functional, reliable, and usable) and a Mac that delivers a particular pleasure for which some people are willing to pay significantly more. Without empathy, Apple wouldn't have been able to anticipate what would make a computer so appealing.

SUMMARY

To deliver value, untethered companies must practice empathy to understand, anticipate, and meet their customers' needs, wants, and desires. Genuine empathy can be difficult for companies that tend to conflate or confuse their own goals with those of their customers. Often, innovation looks like the answer. It certainly can be, but innovation for its own sake often leads to solutions in search of problems. A better solution is customer-centric, empathetic design.

Although people often dismiss design as a "nice to have " or think of it as a finishing touch to make things look nice, empathetic design is really a holistic approach to problem-solving. By centering consumer wants, needs, and desires and using them as its origin point, it enables companies to deliver the full range of user needs and create more functional, reliable, usable, pleasurable, and even delightful products.

It may be counterintuitive, but focusing less on business goals and more on customer wants, needs, and desires returns higher profits. To paraphrase Mr. Costner, if you build it empathetically, they (and the money) will come.

Meeting a need, want, or desire is the only way to produce the value people pay for, and empathizing with the untethered consumer is the only way to understand the value your company can deliver to them. But companies today need to do more than deliver value. They need to deliver it to customers where they are.

MEET THEM WHERE THEY ARE

Pop quiz, Jeopardy-style: What is a podcast?

My guess is your answer was something like, "A popular form of audible content," which is true. They're also perhaps the best example of untethered mobile. To listen to a podcast, all you need is a connection to the internet and a device that can play audio files. You can listen to a podcast through your car, smartwatch, or tablet on the podcast's website or on iTunes, Google Play, Amazon, Stitcher, or a dozen other apps. Podcasts are device-, app- and platform-agnostic.

They're delivered via RSS, an old technology that functions as a simple syndication list and allows podcasts to make their product available to people through whatever app they choose. You can even email or download a podcast as a pure audio file.

Because no single company owns the technology, podcasters create and distribute their material in a decentralized way. This isn't to say there aren't more and less heavily used channels. Most people get their podcasts through Apple or Spotify.

Spotify, particularly, has done a great job of getting its technology widely distributed. It comes pre-installed on some cars and many smartphones. Its ubiquity and business model (free with ads or by subscription without them) make it an easy and attractive option for untethered consumers, but it's still just that—an option. Users can download Overcast, Stitcher, or several other podcast aggregators, and it's up to the creators of a podcast to make their content available or not. There's a tremendous amount of choice. Neither app nor device is predetermined for the creator or listener.

As a likely scenario example for many untethered customers, if you're listening to a podcast at home through your Amazon Alexa speaker and put in your Apple AirPods, the audio switches into your ear. If you walk outside, the podcast goes with you. It will seamlessly take over from your headphones if you get into your car. The podcast not only meets you where you are; it all but follows you around.

BE DEVICE-, APP- AND PLATFORM-AGNOSTIC

In the Introduction, I mentioned companies that thought of their mobile presence as a miniaturized or branded-app version of their websites. There, I was compressing the history of the untethered internet in the context of the horseless carriage syndrome. Here, it's worth taking a deeper dive into the three phases of the untethered internet's evolution. In its first generation, mobile design was largely an afterthought, constraints-oriented and focused on getting a website to work on a smartphone. The second generation was all about developing a branded app that untethered consumers could (and indeed often had to) download if they wanted to interact effectively with a company while away from their computers. Many companies are still living in this generation. Untethered third-generation companies design for mobile first. They frame their entire online strategy around a holistic understanding of their customer's lived experience, focusing on ways they might add value to it. We'll look at each in turn and explore what should stay in the past and what's worth taking forward.

FIRST-GENERATION MOBILE

It's easy to understand why many company leaders were less than thrilled by the advent of internet-enabled smartphones. They'd only recently been convinced that they needed to own a URL and answer emails. Then, just when they'd gone to the trouble of building a website they were proud of, along came these smaller screens that took people away from everything they'd worked so hard to create online.

Rather than thinking about mobile as a means of extending their interactions with users, they saw it as competition. Because people initially made very few purchases on their smartphones, companies didn't see the value of helping consumers engage with their products away from their computers.

This generation of thinking is still around. Recently, I was in the market for landscaping services. I needed someone to come to my house, do some one-time work, and then return on a set schedule to maintain my yard. I got some recommendations from Nextdoor and googled the most highly ranked. Some first-generation sites gave me phone numbers to call. I didn't call. Several had websites that didn't work well on mobile—I had to pinch and zoom and couldn't find a way to request a bid online. On the few sites that provided contact forms, I filled them in. Several of these used the phone number I provided to call me and left voicemails. Three emailed, and two texted, and I set up service with one of them. This required me to create an online account, adding to the ridiculous list of passwords I maintain.

For several months, they came and did the work and mailed me a paper invoice. Unlike many internet natives, I have a checkbook...somewhere. But with most of my bills on autopay, I don't have a routine for handling paper invoices or writing monthly checks. When I got an invoice for five months of unpaid lawn care, it put a divot in that month's budget and an end to my patience. I switched to a contractor app that opens without a password. It lets me schedule weekly or monthly service, skip a week whenever I need to, and pay for it all on my smartphone.

First-generation companies make it inordinately difficult for untethered consumers to contact, contract with, and pay them. This is a more significant loss than they likely realize. In 2020, well over half (61 percent) of visits to US websites were made from mobile devices, and the World Advertising Research Center estimates that almost three-quarters of internet users will use nothing but mobile by 2025.

DON'T CALL US

I always thought my dislike for talking on phones was a function of where I lived. Cell reception is spotty in the city. I can't talk on the train, and if I get a call while I'm in a meeting, I need to step out of the room. But a recent comparison I heard about millennials and phone calls demonstrated to me that it's more than simply a convenience issue. The comparison went like this:

Imagine my friend lives in a mansion, and I decide to drop by and talk to her. When the butler answers the door, I can do one of two things. I can hand him a note and wait in the foyer. The butler will deliver my note without interrupting my friend, and when she finishes doing whatever it is people do in mansions, she will come down to see me. Alternately, I can push past the butler and pace my friend's house, shouting her name until she answers. I can then walk into the room and start talking. To millennials, that's the difference between texting and calling. Because it's asynchronous, a text is less disruptive and more respectful of people's time. It lets them know I'd like to talk to them, but it doesn't demand an immediate response.

> Most untethered technology makes asynchronous communication possible, and most untethered consumers prefer it.

SECOND-GENERATION MOBILE

Second-generation thinking recognizes mobile devices as capable of doing more than providing a smaller, inferior version of a website for those pesky consumers who insist on leaving their computers, but it retains the business-centric focus of the previous generation. This is the domain of owned and operated channels. Second-generation companies build proprietary apps. Believing their products, services, and content need to exist on their platforms with their logos clearly visible, they build branded smartphone apps, ignore other types of technology, and wall themselves off from other businesses and products. They want to "own the eyeballs."

More advanced second-gen companies might produce podcasts or smartwatch apps but require you to log in to their platforms or websites to access them. If a company can't get its logo or name in front of you or get you to its website, it's not interested.

Disney is guilty of this. If I pay for a subscription to watch Disney movies, there's no *customer-centric* reason to make me watch them through Disney's branded app. This strategy prioritizes the short-term gains of getting people onto a platform and exposed to a logo as well as cutting out syndication fees at the cost of limiting people's ability to interact with the company. This sacrifices the loyalty that comes from delivering long-term value.

Rather than meeting customers where they are, owned and operated channels force them to where you are.

Many airlines do the same thing, forcing users to download and install their branded app to manage their reservations, access their boarding passes, or use the in-flight Wi-Fi or entertainment. If you're forcing users to download your app to do things they could just as effectively (and more easily) do over text, in a browser, or through apps they already use for other purposes, you're not putting their needs first.

This doesn't mean there's never a place for a company-branded app if it delivers more value than you could provide otherwise. HotelTonight, for example, began as an app and offers enough of its own value that people are happy to download and install it. Primarily, it pairs unreserved hotel rooms that would otherwise go unoccupied with potential guests who are nearby and happy to get a discount on their room in exchange for the lack of advance planning. People nowadays primarily use it to book hotels well in advance, but it started as a last-minute option for spur-of-the-moment out-of-town travel or for spontaneous "staycations" in the customer's own city. It's a beautiful example of matching a need in the hospitality industry with a corresponding desire in the public and providing value by connecting one to the other.

HotelTonight is clean and easy to use and goes well beyond simply connecting spontaneous people with empty beds. Unlike sites like Expedia or Hotels.com, it presents users with a curated list of upscale to cozy hotels and information about them, along with the ability to automatically request common concierge items (like a toothbrush or extra pillows), and it even offers tips about what to do nearby. A few taps reserves the room. Some of HotelTonight's partner hotels have even streamlined check-in so a user can get their room number ahead of time, go straight there, and unlock it with a code on their smartphone. The hotel gets a body in what would have been an empty bed (and often an additional cocktail in the bar or room service meal). Users get a streamlined, often upscale experience at up to 50 percent off with minimal effort.

Spotify is another app that more than earns the effort it takes to download. Spotify provides users with access to a vast catalog of music, podcasts, and content without requiring them to buy it by the song. It lets you set up playlists and download the ones you want to play when you're offline. It wasn't the first untethered music service; iTunes has been part of Apple's DNA since the days when "music player" meant a separate piece of hardware. When smartphones largely replaced our cumbersome iPod, cell phone, and digital camera collections, iTunes transitioned gracefully. But not universally. Despite entering the market a full five years after iTunes (2006 and 2001, respectively), Spotify caught up with and overtook Apple Music (165 million versus 78 million) because it was more untethered. Spotify is platform-agnostic while Apple Music is not.

THIRD-GENERATION UNTETHERED

Third-generation companies recognize that the untethered internet enables them to engage with a customer anytime, anywhere. They know that building a non-transactional, empathetic relationship with customers means engaging in every way that meets their needs, wants, and desires.

Rather than focusing on what works best for their brands, they look first at their customers' total, lived experiences to understand the ways their needs change by time of day, across a range of locations, and through various activities. Then they look for ways that their products or services might add value across that entire spectrum.

Facebook (although far from perfect in some ways) has a truly a third-generation untethered approach to being device-, app-, and platform-agnostic. It does such an excellent job of meeting people where they are that it's become inseparable from their lives (more on inseparability in Chapter 7). Most untethered consumers don't feel like they log in and out of Facebook—it's just *there*. You can log into other services with it and receive messages from it on your computer, smartphone, or smartwatch.

THIRD-GENERATION STRATEGY

Untethered thinking recognizes that nobody can (or should) design a product so good that people will never again leave their home computers. People lead complex lives that change day by day and hour by hour. They spend time at work, at home, and many places in between. Having identified a value it can provide, an untethered company looks at where and how it can meet people where they are to deliver that value to users across that whole range of experiences.

Rather than deciding which devices to invest in and trying to ensure it gets direct credit, an untethered company looks for ways to build relationships and meet people where they are. It asks, "Where are we uniquely qualified to provide value through our customers' days in a way that benefits us both? What devices would help me do that? How do behaviors change given different times and places?" The company then designs accordingly, mobile first.

MOBILE FIRST

Untethered design strategy begins with mobile and builds its success criteria and KPIs around it. It treats mobile as the primary use case (or "happy path," in designer-speak), working backward from the smaller screens to the larger ones.

The restaurant reservation service OpenTable is a wonderful example of platform-, device, and app-agnostic third-generation untethered thinking that's mobile-first and meets people where they are. It has both a website and an app, but it's also embedded in many individual restaurant sites and integrated with Google and Facebook.

I might search on my car's navigation system for a lunch place near my current location. In just a few clicks, I can reserve a table, get a text confirmation that I can forward to friends, and plot my route there. There's no need to go to the café's website or OpenTable, sign up for anything, or key in my name and contact information. It lets me do exactly what I want to do as efficiently as possible. I can do the same thing from the café's website or Yelp for lunch today or for a special dinner three weeks from now (in which case it will send me a reminder the day of). It fits neatly into my life, making it easier for me to do something I want to do. It also has a branded app that I occasionally use to find new restaurants, but I would never have considered OpenTable for that use case without its base table-booking utility. OpenTable doesn't have an opinion about what device I use or whether I ever visit OpenTable.com. It's focused on making it easy for me to make, change, cancel, remember, and show up for restaurant reservations.

A landscaping service that operated this way might have had an ad on Thumbtack that, with a single click, would text me a bid for weekly maintenance or generate a text bot to ask a few questions I could answer asynchronously. It would let me spec out, schedule, *and pay for* my service in whatever way was easiest for me.

SUMMARY

Meeting people where they are, on any device, app, or platform—even if they don't get credit for it immediately—doesn't come naturally to most businesses. It requires trade-offs and trust, but it allows businesses to deliver more value more seamlessly, keeping them top of mind and making it more likely that untethered consumers will interact with them more frequently.

Untethered companies fundamentally restructure how they think about how consumers interact with them online. Rather than restricting the platforms on which they're available or trying to force people to their websites or apps, they consider how they can add value to their customers' lives wherever they are first. They then determine the device, platform, and app pathways to provide that value.

Meeting people where they are requires companies to be device-, app- and platform-agnostic, but it also requires meeting them *as* they are—adapting what they offer consumers to the amounts of time, attention, and memory they have available.

CHAPTER 4

MEET THEM AS THEY ARE

In 2014, the Giants were in the World Series. I watched the first game at a party with some friends and the second in my apartment while I caught up on laundry. I was presenting at a conference the next week and needed a haircut. The only time I could get one scheduled was during the third game. In a tethered world, this would have been agony, but even eight years ago, the game was untethered.

At home, I was watching and listening to the game through ESPN's live stream on my smartphone. The audio switched to my AirPods when I put them in, and when I got in my car, its sound system picked it up. Even during my haircut, I had options. I could type "Giants" and "World Series" into Google and get the score or follow the written, minute-by-minute updates on ESPN's GameCast. If I'd wanted to wait, I could have gotten an asynchronous play-by-play when I got home.

In its ideal world, Major League Baseball would probably prefer that every fan watch every inning of every game in a stadium wearing their team's jersey and looking away from the field only long enough to get another beer. But it's not the world we live in. MLB recognizes that and, along with ESPN, did a great job providing ways for me to interact with the World Series at different levels of attention, in various places, and in time increments that ranged from several dedicated hours of very nearly full attention to a few second-long glances. At the same time MLB was serving semi-engaged fans like me, its mobile apps were also catering to hardcore fans in some of the most innovative and customized ways out there, but we'll talk about that later.

BE FLEXIBLE

The untethered consumer is always "online" but with different degrees of presence. To reach them, companies need to anticipate and adapt to the varying modes of interaction, degrees of attention, and amounts of time consumers have available.

MODES OF INTERACTION

In the previous chapter, I held up Spotify as a great example of an untethered company that has gotten its technology widely distributed. It's willingly ceded its primary brand presence to meet people where they are. It's integrated with the Nike running app and Strava so I don't have to switch out of one app to access my music and back to track my run. But it goes beyond simply being available. It adjusts to changing physical constraints to remove barriers, meeting users not just where but also as they are.

In my car, my eyes aren't (or shouldn't) be on my phone. Spotify transfers control from the app to my car, enabling me to skip tracks from the steering wheel control panel or the in-dash screen. If I get a call while driving, I don't have to reach for my smartphone to pause my music. Spotify does that for me.

Netflix and YouTube, which are primarily video-watching services, have made the same calculation and found ways to still provide value when users can't watch videos by developing audio-only options. Voice-control interfaces like Siri and Alexa (although not perfect) are another way forward-thinking untethered companies are working to accommodate different modes of interaction. In the same way that providing audio without video allows consumers to continue using Netflix when they're watching the road, voice commands make Google search or shopping on Amazon possible without needing to take your hands off the wheel. Although I am not sure I would recommend shopping while driving, that's a personal decision.

ATTENTION

Human attention isn't a toggle switch; it exists along a continuum. On the far end is complete attention—a person is totally focused on a single input. Full attention is rare, reserved for falling in love or being in a dark movie theater with surround sound after you've finished your popcorn. Simply by virtue of being mobile, untethered interactions really can't capture or hold this kind of attention. Non-stationary people need to devote at least some modicum of attention to remaining non-horizontal.

Watching TV and occasionally checking your phone or listening to a podcast while running occupies the middle of the spectrum. Here, your attention is evenly divided, either switching between inputs or simultaneously attending to more than one. At the furthest distance from complete attention is inattention. Here, you may be sitting in a meeting with your phone in your pocket or have it face-down on the table while talking with someone you're not imminently falling for. You have close to complete attention on something else or are splitting it mostly between other things. However, a level of subconscious attention is still available to your phone.

Mobile gives you access to a user's subconscious level of attention you wouldn't otherwise have. In marketing terms, it allows you to become (and stay) a part of a person's consideration set.

Consideration Sets

A friend of mine illustrates the power of consideration sets brilliantly by asking you to name three brands of ketchup, knowing you'll probably come up with Heinz first and Hunt's second. Most people can't think of a third, and yet Hellmann's, French's, and Annie's Organics all make this most American of condiments. And although you probably couldn't tell the difference between them in a blind taste test, you're much more likely to buy Heinz or Hunt's precisely because those are the brands you think of when you think of ketchup. They are in your mind's ketchup consideration set.

Consideration sets are one of the shortcuts our minds use to manage the enormous (and steadily increasing) amount of information we ask them to process. Rather than catalog everything about everything, our minds maintain shortlists of grouped things, experiences, and concepts. In his book *Brand-Simple*, Allen Adamson describes brands as "shortcuts on your mental desktop." Inclusion is based on familiarity. If I asked you to name three sports, football, baseball, and basketball would be more likely to make your list than hurling, polo, or quidditch. The more you see or hear a brand's logo or name, the more likely it is to make it into your consideration set. Becoming (and staying) part of consumers' consideration sets is the purpose of brand awareness (more on this in Chapter 6).

Our smartphones, smartwatches, and other untethered devices stay present in our unconscious levels of attention, and their alerts and other notifications exert an almost irresistible draw on our attention. (If you doubt me, resolve to wait five minutes before checking your phone the next time it sends one and see how able you are to resist.) They accomplish this by hacking the brain's use of consideration sets. We maintain them in our unconscious attention because they might, at any point, notify us of an incoming message. When they do, our brains try their usual shortcut. But the consideration set of what that buzz, bell, or bing might be is almost infinite. The shortcut isn't short. It's complex, and that makes it compelling.

Rather than taking advantage of this access to the shallow end of the attention spectrum, many companies make the mistake of aiming for the opposite, near-total side. They believe attention is a choice people make and think, *What do I have to do to make them pay attention?* Innovative untethered companies take advantage of the attention spectrum and have a subtler approach. They look at how much attention people have available at different times and in different places and adjust their offerings accordingly. They know that every touch point increases familiarity and helps locate and maintain their brand in consumer consideration sets.

Variable Attention Touch Points

Because people tend to give something very close to their complete attention to video games, it's understandable that the companies that produce them have been among the slowest to capitalize on the possibilities of untethered technology. Some games have refused to make any portion of their environments available beyond the PC or console. But they're missing out on opportunities like audio content and short stories set in game worlds. *World of Warcraft* did exactly this in the lead-up to its *Legion* expansion with *The Tomb of Sargeras* audio drama, as did *The Callisto Protocol* with its *Helix Station* audio prequel in advance of its game release, even going so far as to cast big stars like Gwendoline Christie and Michael Ironside. These types of content have enormous potential to be huge hits and provide multiple touch points between gamers and their games.

Audio-only content (and particularly podcasts) is well-suited to the untethered consumer's fluctuating attention. Because this content tends to be less information-dense than audiobooks, people can drift in and out of the information flow of a podcast while they drive or grocery shop. Companies that aren't able or willing to provide customers with ways of interacting with their products or services at reduced levels of attention will find that attention turns to other companies. We can look again to Major League Baseball as a company that understands this potential. It offers a monthly subscription service called MLBAudio, which offers live broadcast, on-demand pitch-by-pitch announcing, team-by-team highlights, and game recap audio for every game played in the season—all available in multiple languages. The MLBAudio service also offers Amazon Alexa and Kindle integrations to make sure that if you have a sliver of time, you can choose to fill it with baseball.

Attention and Memory

Attention and memory are deeply connected. The more attention you give something, the more likely you are to remember it, and the same things that disrupt attention disrupt memory. Because people rarely give their complete attention to their untethered devices and because they're more likely to be interrupted while using them than they are while seated in front of their computers, effective untethered design makes two psychology-based accommodations to users' memory. The first is based on the concept of cognitive load, and the second on the primacy and recency effect.

Cognitive load conjures an image of the mind as a wheelbarrow—you can only pile so many apples in before some start to roll off—but I tend to think of it as the brain's amount of available processing power. The classic experiment goes like this: scientists ask a subject to remember a string of numbers or unrelated words and then to perform a series of simple tasks like writing a few sentences about what they did over the weekend or picking between a selection of snacks. Even with reminders to remember their assigned numbers or words, it doesn't take many additional activities to overburden subjects' working memory. Their cognitive load is exceeded, and the items held in memory roll off.

The implications for mobile design are straightforward: don't ask users to remember things. Any information your untethered customers will need moving through a process should be put on the screen and kept visible until they no longer need it.

Another interesting trick of human memory is remembering most vividly the first and the last parts of an experience and all but forgetting the middle. This is called the primacy (first) and recency (last or most recent) effect. If the last part of your excellent dinner at a nice restaurant is having to pay for it, your memory of the entire experience will be better if the bill comes with a mint. If my first encounter with in-flight entertainment is being railroaded into downloading the airline's app, my negative initial experience discounts my enjoyment of the movie. Thoughtfully designed untethered experiences account for this aspect of memory and allow users to enter them seamlessly and leave them with a good taste in their mouths.

TIME

People rarely (intentionally) sit down for an hourlong session with an untethered device the way they might in front of a TV or computer monitor. Even when they are being intentional about completing a particular task, they're likely to be interrupted. Whether this interruption comes from needing to head into a meeting or the arrival of a new text alert, people swipe out of apps or between them frequently.

Most banking apps do a good job of recognizing this and are designed for brief interactions. They know you're standing in line at the store and want to quickly check your checking account balance before you stick your debit card in the machine. They let you log in quickly and put your most frequently used accounts on the first screen that comes up. Sure, they could try to sell you on opening a retirement account or present you with the option to order new checks, but they recognize the most common use case and provide for it, prioritizing your needs and meeting them efficiently.

Unfortunately, not all companies are equally aware of an untethered user's fragmented time. Imagine you're waiting for a friend to join you at lunch and remember you need to book a flight for next month. You go to Expedia or Kayak and enter your dates and destination. After a bit of comparison shopping, you pick a flight, enter your passenger information, and move on to the payment screen. Then your friend texts. He's forgotten the restaurant's address, so you swipe over and text it to him. You go back to pay for your flight and find yourself back on the home screen with blanks in the dates and destination fields. This is maddening!

Because long, single-page scrolling screens make heavy demands on a device's memory, mobile design tends to break a process into multiple single screens. You hit "next" rather than swipe up. When you background an application, it is likely to dump the information, forcing you to start over. But this is a design choice (or lack thereof). Savvy designers expect users to be interrupted at any point and save the user's place in the process flow.

Modern mobile versions of games are notorious for failing to plan for interruptions. For example, while playing Fortnite's mobile version, if you did so much as reply to a text, it disconnected you from the game. It *really* wanted you to sit in front of your computer to play. If you weren't going to give it your complete attention, it didn't really want you interacting at all. Sure, allowing you to play on a smartphone is a modern marvel, but by clinging to a desktop's style of interaction and assuming the same level of attention, the "miracle" often doesn't deliver in the situations players most find themselves wanting it to.

Tesla, in contrast, looked at the fractured time, incomplete attention, and variable modes of interaction of untethered technology and decided it was a fine place to sell you a car or even operate it. A Tesla is a considerable investment, so it would be logical to assume it's one that people will devote quite a lot of time and attention to making. It's not unreasonable to think nobody would buy a car on their smartphone, but about 30 percent of Tesla's customers do. These aren't impulse buys. Tesla simply lowered the barriers to purchasing its cars by being flexible; anticipating their customer's needs, wants, and desires; and meeting the customer where and as they were. Yes, Tesla has also reinvented the dealership system, but it still starts with having an untethered purchase experience.

SUMMARY

The untethered consumer's needs, wants, and desires change depending on where and how they are. Untethered design recognizes that although a person might not be in a place where they can open an app or have the attention to interact more than by passively listening, a company can still deliver value to them. The more flexibly a company provides value across the spectrum of time, attention, and interaction modes, the more opportunities it has to do so.

Limited modes of interaction, partial and fluctuating attention, and small increments of interrupted time can seem like constraints. To a certain extent, they are. They're also at the core of untethered technology's unique power. Because it is more personal and holds users' attention less completely, untethered technology attracts users' attention more magnetically. The key to unlocking this power to compel is relevancy.

HAVE SOMETHING THEY WANT

I got a text notification from two different companies this morning. The content was similar—they both told me about a sale that was active until five o'clock, but one irritated me, and one changed what I ate for lunch.

The first was from World Market. They'd gotten my phone number when I signed up for a sweet discount on tea, and the push notification suggested that outdoor dining sets were on sale. I didn't even have to pick up my smartphone to know the other was from Taco Bell. Its cool, custom bell sound identified it, immediately reducing my sense of urgency about (and thus the interruption of) checking my phone. My favorite taco—the Fiery Doritos Locos Taco (of course)—was buy one, get one free, and I could order now for pickup later. I didn't, but on my way to run another errand, I stopped and got lunch.

Taco Bell could easily have been forgiven for thinking the untethered internet didn't have much to offer it. Most chain fast-food restaurants stick to TV and website ads and count on unrealistic photos to make your mouth water. People are going to get tacos when they want tacos, right? No need for a mobile app. But Taco Bell has been much more creative, finding ways of using the untethered internet to both maintain its food in consumers' consideration sets and keep consumers interacting with its brand. It's what World Market was probably trying to do as well, but I unsubscribed and went to Taco Bell for a reason—relevance.

BE RELEVANT

Companies that want to leverage the untethered internet's unique ability to foster non-transactional relationships with consumers need to ensure their communications are relevant. Relevancy is based on an empathetic understanding of customers' needs, wants, and desires. It allows you to meet them where and as they are in ways they welcome rather than resent. This requires getting users' permission to engage with them, properly timing that engagement, and then actually being engaging.

PERMISSION

The first difference between my alerts from Taco Bell and World Market was that by downloading Taco Bell's app, I'd initiated an interaction with the company that gave it explicit permission to ping my smartphone. While some people might argue that it's problematic for a company to trigger me to think about tacos when I wasn't choosing to, Taco Bell's chain of consent kept its text from feeling like an imposition or unwelcome intrusion. Not only had I downloaded its app, but I'd authorized push notifications. I'd functionally "told" Taco Bell I didn't mind being prompted occasionally to consider saving some money on a taco, and its settings let me opt in or out of specific notification types.

What's funny to me about my own behavior is that tacos aren't expensive. The money savings I get when I buy tacos I might not otherwise have purchased is, at most, a buck or two. I would never pick up the phone and say, "Hello, Taco Bell! Please tell me when I can have a taco," but I don't mind when it does precisely that. I know I've given my permission in exchange for a discount, and Taco Bell doesn't abuse it by texting me every day.

Customer-initiated contact comes with permission to respond but doesn't necessarily imply the okay to do more. If a customer emails with a question, don't automatically swipe their email for your mailing list. If they ask to be notified with updates on an out-of-stock item or the status of an order, don't assume permission to advertise to them through the same channel unless you explicitly ask.

I signed up to have my pharmacy notify me when my pre-scriptions have been filled. I'm happy to have them remind me when it's time to reorder—or even to get my flu shot—but I don't want CVS to text me about two-for-one photo prints. Permission is domain-specific.

Even within a domain where a customer has given their permission to be contacted, it's not an on-off switch. Major League Baseball understands that you having interacted with them doesn't necessarily mean you want all baseball, all the time. When you select your favorite team in its app, you are also presented with a screen full of granular controls for noti-fications at the league and team level: general news, game start times, lead or score changes, pitching changes, condensed game summaries, and availability of video highlights. MLB could easily have signed you up for every type of notification by default and buried those controls in some hard-to-find menu. Instead, it built goodwill by asking permission and establishing relevance first. Of course, hardcore fans can choose to eat, sleep, and breathe baseball alerts, but that's their prerogative that MLB respects.

> If you don't have permission, you're not relevant.

TIMING

Although all but the most inept spammers have learned not to send push notifications when they're likely to wake people up, few have fully leveraged the untethered internet's ability to sensitively time them. An alert on a two-for-one muffin with my coffee purchase that arrives at dinnertime isn't relevant. On the other hand, if I often get a coffee and pastry for lunch, noon would be a better time to let me know about the deal than closer to breakfast.

Frequency is another critical element of timing. There's almost no company I want to hear from every day, and I know I've unsubscribed from lists I would have been happy to hear from occasionally only because they messaged me too frequently.

In a great example of understanding the subtleties of good timing, Apple's iPhone has long offered a "do not disturb" mode that allows only certain texts or calls to reach the customer. More recently, it's evolved into a feature set called "focus," which includes smart defaults for sleep, driving, personal, and work and can be shared across all connected Apple devices. The focus functionality lets users select which people or individual apps are allowed to penetrate their focus and includes options for creating auto-reply messages and turning on features like dark or low-battery mode while indicating to others that notifications are turned off.

The newer modes like sleep and driving have even more precise functions. Driving mode can be triggered automatically when the smartphone or smartwatch detects that a car has connected to its Bluetooth or when its accelerometer indicates you're moving faster than the average human can run. To turn it off, the driver has to respond to a prompt that explicitly says, "I am not driving." Sleep mode can start a customizable "wind-down" function that starts a half hour before your scheduled bedtime to start reducing the apps that might tempt you to stay up late and offers planning tools that help users hit their sleep goals. Android offers a similar suite of tools called digital well-being. In both cases, this functionality is freely available on the devices. It's not a subscription or revenue driver (although it's likely there is some liability coverage involved). Apple and Android understand that their customers will be happier and more engaged if their interactions with people, apps, products, and companies are welcome and properly timed.

ENGAGEMENT

While it might seem counterintuitive for a fast-food company to invest in a robust device-agnostic mobile app, Taco Bell clearly demonstrates how much more can be done with one than with a simple mobile-enabled website that lets people order a taco from their smartphone.

An app lets Taco Bell get to know me—or the Doritos-Locos-Tacos, never-as-a-combo-meal, often-before-five-o'clock, taco-buying me. It doesn't send me combo-meal deals, and it doesn't let me know when I can get 10 percent off a Burrito Supreme because I don't like those products.

> Even if you have permission and your offer is well-timed, if it's an offer of something a person doesn't value, it's not relevant.

To extend our analogy from the previous chapter, if I've never been invited to my friend's mansion and she's never given me her address, it would be creepy if I showed up on her doorstep, even with a calling card. If my timing is bad—if I arrive uninvited when she's at work or sitting down for dinner with her family, it's an unwelcome intrusion. And even if I have permission to drop by from time to time, I don't want to arrive with flowers she's allergic to or a box of chocolates when she's just started a three-day fast.

Ultimately, the difference between spam and relevant points of contact is mindset. The first comes from asking the wrong question: *How do I get people to open my app, go to my website, or buy my product?* The second starts with: *How can I offer someone something of value at a time and in a way they'd appreciate?*

PRIVACY AND RELEVANCE

Most untethered consumers have fluid relationships with their online privacy. Knowing that there's no escape from advertising, I'm willing to accept more transparency and less privacy to increase the relevancy of the advertising I see. I'd rather be well-targeted by marketers and learn about new green technology products and the kind of coffee I enjoy. I believe that as consumers, we're better off allowing companies access to certain portions of our data because they use it to improve our experience. But I recognize the trade-off.

There's no way for products to design themselves around a person's life if they don't have access to information about them. Choosing to share your habits and data makes some remarkable and life-enhancing things possible, and I'm willing to sacrifice some privacy for a better experience. But as tech-forward and untethered as I am, even I have limits. I strongly recommend companies have them as well. The desire for increased relevance is admirable, but there is such a thing as too much.

In my opinion, companies make a mistake when they impose opt-out rather than opt-in privacy policies. I understand the impulse, but it runs counter to the untethered philosophy. People's more personal relationship with this technology makes having an invitation more important. Building a relationship with your customers requires empathetically accounting for their privacy goals in addition to your business ones. No, you can't meet people where and as they are if they don't opt in to sharing their information with you, but if you simply assume it's okay for you to take and use it unless they expressly opt out, you're undermining not only your relevance but your brand.

BRAND AWARENESS

It's an old saw that 100 percent of the people who don't know about your product won't buy it. Certainly, the untethered internet's ability to create multiple small touch points with consumers makes it an appealing way to bring a brand to mind. Doing this poorly, though, is worse than nothing at all. I got irrelevant texts from World Market, and each time they irritated me. I finally invested the two seconds it takes to unsubscribe. Until I did, each point of contact caused me some mild irritation I associated with the brand. Will it keep me from going into one of its stores again? Probably not. But the net net of their investment was negative.

Taco Bell, in contrast, has benefited both directly—it's absolutely gained a few additional taco sales—and indirectly. Even if I never read the app's notifications, the Taco Bell brand is brought to my awareness each time I hear its distinctive alert sound. It keeps the brand active in my consideration set. I may think, *Oh, that's the Taco Bell app. I don't want tacos today*, but I've thought, *Taco Bell*. Too few designers consider engaging with users on this end of the attention spectrum, focusing exclusively on what happens once users have engaged with the app directly or are already on the website.

Brand awareness can also work in even subtler ways. During the COVID-19 pandemic, Scotts Miracle-Gro was looking for a way to connect with new home gardeners, many of whom needed education, inspiration, and human connection. Scotts worked with the creative agency Banter to craft a Miracle-Gro-branded podcast called *Humans Growing Stuff*. This wasn't just classic product placement of the "brought to you by Ovaltine" sort. The podcast enlisted a passionate host and invited various unexpected celebrity guests (like comedian Jim Gaffigan, actress Lauren Conrad, and chef Vivian Howard) to share their gardening stories. The podcast's icon doesn't include the Miracle-Gro logo, and the company is rarely mentioned in the podcast itself. This might seem counterproductive for its business goals, but people are smart and know when they're getting a long-form ad in the guise of a podcast (or book). *Humans Growing Stuff* delivers genuinely useful content, including gardening tips and personal stories, and even tackles some surprisingly heavy topics like toxic masculinity, representation, and mental health. Miracle-Gro was certainly being non-transactional but because it also empathized with people who'd just picked up gardening while stuck at home, it understood an untethered mindset was the best way to reach them and was able to give them something relevant and engaging that they genuinely wanted or even needed.

SUMMARY

Relevancy, like so much else, comes back to anticipating needs. This can be as unsubtle as asking customers whether they want to hear from you or as nuanced as watching how they interact with your offerings and tailoring your outreach accordingly to provide the right offer at the right time.

The more relevant an offer is, the more effective it's likely to be, and perhaps the best way to make something relevant is to allow users to customize it to their individual needs, wants, and desires.

HELP THEM MAKE IT THEIRS

Once upon a time, when you walked into a Blockbuster store, you were greeted with blockbusters even if you only ever rented tapes (yes, actual, physical video cassettes) from their tiny foreign film section. If you wanted a recommendation and were lucky enough to find an employee who wasn't stuck behind the counter, and if you were willing to explain your taste in general and what you were looking for that night in particular, your odds of getting something you enjoyed were still pretty slim. Oh, and there was no way to watch a preview first.

When it came time to check out, you produced your laminated Blockbuster membership card, of which you were allowed to have no more than three, with everyone on that card required to live at the same physical address. If you were married with two kids old enough to want their own cards, you were out of luck.

Of course, some of these limitations were due to the physical constraints imposed by available technology. Still, part of Blockbuster's many successors' success comes from their willingness to let go of that company's level of control and create individual one-to-one relationships with multiple users on the same account.

Far from the one-store-for-all Blockbuster days, there is no longer a core experience for Netflix. Every person who logs in is greeted with a variation based on their tastes, activity, and location. Multiple people can share one account, and one person can create multiple profiles. The same family of four that shared three Blockbuster cards could have a Netflix profile for each family member plus one for the kind of movies they like to watch together. Customizable profiles let people maintain their own histories and create their own wish lists. They also allow Netflix to personalize each person's experience by showing them a home screen populated with the shows they've recently watched. These personalized home screens also offer algorithm-tailored recommendations, which generally do a very good job of anticipating what a specific user might enjoy next, despite the occasional semantic fail (after I watched *Ten Years a Slave*, Netflix helpfully suggested *Roots* and *Django Unchained* because "We noticed you like slavery").

BE PERSONALIZED AND CUSTOMIZABLE

In Chapter 3, I joked that Netflix would probably let you watch a movie on your microwave if it had a screen, but the company's device agnosticism points to an essential link between that untethered attribute and customizability. Very few people have the same personal relationship with their televisions that they have with their smartphones because the relationship lives in the profile, not the device. I don't mind telling Netflix who's watching, but I expect the Starbucks app on my smartphone to know who's ordering coffee, which drinks I order most, and how many stars I've collected.

Whether personalization and customization result from the one-to-one relationship people have with their untethered devices or from an individual profile on a shared device, the untethered consumer expects to customize their experience to suit their unique usage and tastes and to have it personalized for them. When done well, personalization and customization make the most of incentive programs, increase user agency, support wish lists or playlists, and deliver a better, more seamless experience.

Blockbuster may have been an extreme example, but many companies carried over the storefront mentality from in-person to online. There are e-commerce sites I frequent that, no matter how often I've shopped there, still greet me with a standard front page showing me the same range of choices anyone else sees. Even Amazon, which has some customization (shopping history, wish lists, and account information), has failed to fully capitalize on personalization. The first page I see includes some "You recently browsed" items below the fold. However, it still splashes an upcoming sale or the latest release from Amazon Studios across the initial screen, no matter how little it may have to do with me personally. It still feels like walking into a big-box store. Amazon is also fantastic at recommending items I've already bought.

In contrast, my bank's website doesn't feel like a bank lobby. The entire experience is personalized with my account balances, my savings plans, and buttons to pay bills or transfer funds. The Starbucks app offers much the same level of personalization. When I hit the mermaid icon on my smartphone, the app shows me my regular store, suggests the closest one, and lists my favorite drinks so I can easily reorder. It shows me how many stars I have collected and only asks for a password when I want it to connect to my bank and top up my balance.

Major League Baseball gets personalization and customization right for its most hardcore fans. When you install the MLB mobile app, it prompts you to select your top baseball teams and asks you to select a single favorite team. For me, it's the Atlanta Braves. Sure, I root for the San Francisco Giants and Los Angeles Angels, but I am a Braves fan at heart. The app configures itself with the Braves logo at the top and updates its colors to the team's classic navy blue and red. Now when I open the app, everything I see is related to highlights, stories, stats, merch, rosters, and schedules for the Braves. It even goes so far as to offer an option where I can change the MLB app's home screen icon to the Braves logo.

INCENTIVES

Starbucks' stars, like Panera's reward program, provide a surprisingly strong incentive for consumers. Companies know it's well worth giving away the occasional free coffee or cinnamon bun, but many still require you to dig through your wallet for a punch card or key in your phone number (this is where my trouble with World Market started). Companies like Starbucks and Panera that make it easier for me by automatically tracking my points derive more loyalty from their loyalty programs—and, if I'm at all typical, more income as well. I know I've bought coffees for friends waiting in line with me because I wanted the additional stars for myself!

AGENCY

The appeal of customization and personalization goes beyond convenience to tap into a core human need—the need for control. The psychological concept of agency is a bit more sophisticated than "You can't tell me what to do," but that impulse is at its core. People are happier when they feel able to direct the outcome of events, behave with autonomy, and choose their own goals. In short, they control their own lives. Or, per Wikipedia, "Agents are goal-directed entities that are able to monitor their environment to select and perform efficient means-ends actions that are available in a given situation to achieve an intended goal. Agency, therefore, implies the ability to perceive and to change the environment of the agent."

The same is true in the untethered environment. Interfaces or process flows that are rigid and make it difficult for users to accomplish their goals frustrate them quickly. Increasing a user's sense of agency can be as simple as giving them the option to set preferences, change the color or background, or personalize their avatar. Allowing people to participate in tailoring their untethered ecosystems not only improves their experience, but it increases their investment in and loyalty to it.

WISH LISTS, PLAYLISTS, AND FAVORITES

Some version of these forms of personalization has been around since the department store wedding registry days. They're all but table stakes for online retail and content delivery companies today, but they exist along a continuum of sophistication. Their most basic function is as simple bookmarking tools that allow users to save a list of things they want or use frequently and make that list available across devices and platforms. A list saved on your smartphone should be accessible on your laptop or smartwatch. A slightly more sophisticated list would give users the option to share these lists or make them private.

Untethered companies do more than simply allow for and store these lists. They use them to increase their understanding of the customers and increase the value they offer. Amazon, for example, lets people save items to a wish list and then purchase them from their smartphone, smartwatch, or Alexa. They can even establish a cadence at which their favorite purchases will be reordered without requiring any additional action. Companies like Thrive Market even use this model to help customers save money. Amazon also lets people set up multiple such lists and make some shareable or even public and searchable for anyone who'd like to send you a present. Try a Secret Santa powered by Amazon wish lists—it's much more powerful than a list of interests.

Other retailers, such as Steam, take such lists a step further and notify customers when an item in a wish list goes on sale. Netflix has gotten so good at this that it's gone beyond recommendations to creating content matched to its customers' tastes.

Spotify goes still further. It allows users to set up multiple playlists, and because it recognizes that most people want to discover new music, it uses those lists to recommend new songs and artists. It even generates entire playlists, telling users which of their existing playlists the new ones are based on. Some lists guess at a genre of music the listener enjoys, while others are things like the listener's top twenty-five songs of the year.

Spotify doesn't limit its recommendations to music. If you use the app to listen to podcasts, it will suggest new ones in a similar vein. If you frequently use it in conjunction with a run-tracking app, it will formulate a playlist that caters to your overall musical taste and with a beats-per-minute ratio that is ideal for running. It's even integrated with the Headspace app to provide meditation playlists. It takes users' customizations and proactively personalizes itself further. Spotify works hard to understand how audio content fits into users' lives and to anticipate their needs. The company's untethered mindset enables it to build a better product.

ADVANCED CUSTOMIZATION

If fast food companies can be forgiven for failing to universally recognize (as Taco Bell did) the possibilities that untethered connectivity offers their industry, car manufacturers have an even better excuse. Cars may be deeply personal, but they don't come off the assembly line that way. A few makes or models may convey volumes about the kind of person driving them, but those profiles are, necessarily, generalizations about swaths of the population.

The first significant new American car company since Jeep's entry in 1941 (or more obscurely, DeLorean in 1975), Tesla is unlikely to dominate the industry.[1] However, in the electric vehicle space, every other manufacturer is struggling to compete with the newcomer. Arguably, Tesla isn't even a car company; it's a technology company that makes cars, batteries, and solar panels. Regardless, its success exposed other car manufacturers' lack of software and connected technology and left them scrambling to follow suit and update their approach to compete in the EV market.

---

[1] As a point of interest, Dodge was founded in 1900, GMC in 1901, Cadillac a year later in 1902, Buick and Ford in 1903, Chevrolet in 1911, and relative newcomer Chrysler in 1925.

Tesla's heavy and early investment in connected technology notwithstanding, it doesn't dominate the electric car market because it built the first or most affordable electric car. It's winning primarily because it understood how it needed to fit into people's lives for an electric vehicle to be viable. From advancing battery technology to software and mobile connectivity, Tesla understood that what people would be buying wasn't the car but the overall connected experience. Smooth steering and zero to sixty miles per hour in 2.4 seconds certainly don't hurt, even if they're not what makes a vehicle valuable day to day for most people.

And perhaps no company more completely embodies the untethered mindset. Taking advantage of people's one-to-one relationship with their smartphones, it treats users' phones as their car keys. Tesla drivers don't even have to log into the app. They just open it on their smartwatch or smartphone, and their car "recognizes" them. This not only makes getting in the car more effortless, but it allows Tesla to personalize each car to its driver.

Both my wife and I drive our Tesla, but the car recognizes which of our smartphones has unlocked it and adjusts accordingly. It repositions the steering wheel, mirrors, and seat and calibrates the music. It makes me feel like my car knows me, and that contributes to my loyalty to the brand. It also contributes to the delight we'll discuss in Chapter 8.

SUMMARY

The more personal (and personalized) a consumer's interaction with a product, the more they feel it's an integral part of their untethered internet experience, and the more secure that company's business is.

Personalization and customization provide incentives to users and increase their sense of agency. They also give companies more information about their customers—information they can use to deliver more value.

A more customized experience is a more enjoyable one, and untethered consumers, like most humans, are more likely to repeat good experiences.

A few companies have taken this integration to the point of becoming inseparable from people's lives.

······· CHAPTER 7 ·······

INTEGRATE WITH THEIR LIVES

Back in Chapter 3, while praising OpenTable's device and platform agnosticism, I mentioned the freedom to use the app without having an OpenTable account. I can interact with it exclusively through third parties to make a reservation at a nearby restaurant (or to find the name of that great Italian place we went to back in June when I was visiting New York). Because OpenTable is integrated with other untethered companies, it's become an inseparable part of my online life.

BE INSEPARABLE FROM PEOPLE'S LIVES

When I'm keeping up with the lives of far-flung family and friends, I know I'm using Facebook, but I almost never have to sign in. The news feed that greets me is both customizable through preferences and personalized by a sophisticated algorithm that pays attention to which posts I comment on and videos I watch. But when I use Instagram to look at my friends' pictures or WhatsApp to message them, I'm still using a Face-book company.

In fact, my interactions with Facebook go well beyond those I actively participate in. Because even if I haven't deliberately logged into Facebook, it's almost always running and connected to me in the background, so it can track where I go online and what I do. While some people find this troubling, I can appreciate the upside. Companies are going to advertise. That's inescapable. By gathering so much data about me and using it to target the ads it shows me, Facebook ensures I see fewer ads that are irrelevant to me. I wouldn't say I'm happy to be advertised to, but I'm happier seeing ads for Tesla and Nike than I would be getting ads for American beer and fast fashion.

Facebook's popularity has declined, particularly among younger untethered consumers who overwhelmingly prefer Instagram, TikTok, Snapchat, and other new-coming apps. But because of the way its structural matrix is configured behind the scenes, consumers still get value from its software. Simply, Facebook can do this because it's identity-oriented, not account-oriented.

MOBILE IDENTITY

An untethered company recognizes that its customers are more than their interactions with any single business—even ones as large as Facebook, Amazon, Google, and Apple. Our mobile identities comprise all our relevant personal and behavioral information and connect the dots between them. My restaurant reservations are a part of my mobile identity because OpenTable allows me to make them through my Facebook or Google profile. Those companies then attach my activity on OpenTable to "me." My mobile identity traverses the online booking and the physical sitting down and eating as seamlessly as I do.

For a company, the necessary mindset shift is from the accounts customers have with you to the relationship between your company and your customers' mobile identities. Savvy, small online retailers have a tremendous opportunity to demonstrate empathy for their customers by adopting this mindset. By recognizing and interacting with our mobile identities, these companies make it easier for us to give them our money.

Direct-to-consumer (DTC) companies like Shopify help make this happen. Shopify isn't a retailer but a platform for small to medium-sized companies that makes it easier for consumers to make purchases. You have likely purchased through Shopify without even knowing it. If you want to spot it in the future, Shopify's checkout screen has a distinctive right-side column with your items listed and typically offers a variety of express payment options, including Amazon Pay, Google Pay, PayPal, and Apple Pay.

As an example, several months ago, I wanted to hang some lights on my covered patio and ended up in an obscure corner of the internet searching for ones that fit the size and shape I needed. There, I discovered Alumahangers. I'd never heard of the company before and didn't have any kind of order history with it, but I didn't have to enter billing and shipping addresses or give the company my credit card number to make a purchase.

Instead, I bought my Alumahangers through Amazon. Not *on* Amazon, but through my Amazon account. Rather than needing to trust whatever unknown credit card processing service such a small company might use, and without having to create yet another online account with a matching password I'd then need to keep track of, I placed my order with a single click through a company I already trust to handle my data.

This is good for customers and small businesses, but it's good for Amazon too. The more a person does with and through the portion of their mobile identity that Amazon maintains, the more inseparable Amazon becomes from their life.

On the back end, Alumahangers almost certainly created an account on my behalf, but because that account is connected with my overall mobile identity rather than being isolated and off on its own, it's an advantage rather than a hassle. Months later, visiting my parents in Atlanta, I told them about the lights I'd been able to hang with these great patio hooks. Then with a tap and a swipe, I was able to re-find Alumahangers, swap my shipping address with my parents', and dispatch hooks to them from their patio.

Your mobile identity extends beyond your payment methods and billing and shipping address. Through it, you can connect your email to your calendar software or your Slack text messages to your Alexa. Companies like Zapier and IFTTT (If This Then That) have built entire businesses around connecting business applications (Zapier) or consumer technology products (IFTTT) to each other in unexpected ways.

IFTTT

IFTTT operates by creating triggers and actions that integrate with various products. For example, a smart home security system like Ring or Vivint can generate a trigger when you lock the front door. This is the "if" piece that triggers an action like telling your Phillips Hue lights to turn off (*if* I lock my front door, *then* turn the lights out).

In 2017, Domino's introduced an IFTTT integration that allowed customers to do crazy things like set triggers and actions such that if Fitbit recorded you'd reached your daily calorie goal, it would order your favorite Domino's pizza and then turn your Phillips Hue bulbs green when the Domino's Pizza Tracker indicated your pie was on its way. While there's more novelty than practicality in such a feature, it demonstrates that Domino's has an untethered mindset, which quite literally took the company from near bankruptcy to being one of the most innovative tech companies in the restaurant business.

It all started In 2011, when Domino's CEO Patrick Doyle challenged his IT team to "make it so a customer could order a pizza while waiting for a stoplight."[2] This approach birthed powerful revenue-generating product features like Easy Order (which underlies the IFTTT integration) and the infamous Domino's Pizza Tracker (which allows the stoplight customer to know if they can beat the pizza home). By connecting with customers' mobile identities, Domino's was able to offer them multiple ways to order a pizza and thus integrate the company with their lives.

Extending a person's mobile identity to proxies such as their home or other location has spawned an industry segment called "smart home automation." You are likely familiar with Nest thermostats and Ring doorbells. These are products that exist in a physical location (typically a home) and enable remote interface with them through websites or mobile apps. You can crank up your home AC from your car on your commute back from the office or see who's at the front door without walking downstairs.

Once companies realize that from the untethered perspective, location is just a variable in a customer's mobile identity,

---

2   Jonathan Maze, "How Domino's Became a Tech Company," Nation's Restaurant News. March 29, 2016. https://www.nrn.com/technology/how-domino-s-became-tech-company.

exciting applications emerge. Philips Hue offers a standout example. Phillips is a major manufacturer of commodity electronics like light bulbs (not necessarily an obvious arena for untethered internet innovation), and colored light bulbs are certainly nothing new. But when you buy a Hue bulb, you get a powerful untethered platform that allows you to control your lights from anywhere and sync them with the movie playing on your TV or the music playing through your Spotify account. Hue bulbs are integrated with IFTTT, Amazon Alexa, Samsung SmartThings, Logitech, Vivint Security, and many other companies. This means that thousands of products you already use can automatically turn your lights on or off, change their color, and set entire moods or scenes.

Phillips started its integration program with single bulbs but has expanded the technology into light strips, outdoor spotlights, holiday lights, and even projectors inspiring the creation of smart furniture like tables and bookshelves so that they, too, can contribute to the Hue scene change. Today, streamers use it to create amazing on-camera spaces, and it spices up lawn parties and barbecues and gives a modern twist to classic holiday house decorations all because Phillips knows that no matter where you are, your "identity" can be your home, your office, your yard, or just a special room.

Perhaps no company has done a better job of integrating itself seamlessly into people's lives than Spotify. It seems to have made its job to integrate with any device you might use while listening to anything. It syncs seamlessly with most cars and smartwatches and with activity-specific apps like Nike's Run Club app, the Waze navigation app, and the meditation and sleep app Calm. It not only stores your playlists but makes celebrity playlists available to you, creates new ones it thinks you might like, and tailors these new playlists to different use cases—one for running and another for relaxation or sleep. And because it houses millions of tracks and podcast titles, it's unlikely you'll need to go elsewhere for anything you want to listen to.

Although Apple basically invented digital-only music with the iPod, its reluctance to move to a subscription model, to become platform- and device-agnostic, and to widely integrate ended up costing it its lead. Spotify now has more than twice the number of subscribers. Even though Apple Music also makes you buy all the songs, I actually used to buy the specific ones I loved running to because I needed the Apple integration to work on my running apps. I also used to download podcasts through Apple in order to get them on my watch. As soon as Nike and Strava introduced the Spotify integration, Apple Music lost me. It was never likely to reach customers who use Android devices (which is a majority of mobile users), but the fact that Spotify is effectively stealing people like me who have Apple Music built onto their devices is proof of the power of Spotify's more untethered strategy.

SUMMARY

When an untethered company becomes part of the substrate of a consumer's mobile identity, it becomes inseparably integrated with their life. This, in turn, increases both the company's value to consumers and consumers' value to the company.

With trust and empathy and a focus on meeting consumer needs, wants, and desires where and as they are across platforms, devices, and apps, companies are in a unique position to deliver not just value but delight.

DELIGHTFUL!

The other day, I was editing family photos. I got a little shot of dopamine with each one I finished retouching. Then my power went out. In Chapter 4, I talked about reducing cognitive load by frequently saving user input so people can enjoy making progress on tasks like booking a vacation when they have small increments of time rather than forcing them to be in one place and complete the whole job in a sitting. This creates value for users on a couple of fronts. We're biologically programmed to enjoy the feeling of making progress toward a goal. The role of dopamine in our internal human survival wiring is something Simon Sinek speaks about at length in his book *Leaders Eat Last*.

If making progress feels better than standing still, losing progress (or having it "dumped," to use my mom's jargon) feels even worse. Recognizing this, the best software UX includes the near-constant background saving of files to the internet or internal cache and automatic recovery. Still, when Photoshop popped up the message that it had recovered four unsaved files, I was delighted.

BE DELIGHTFUL

While no product or service ever has to be delightful, going that small step beyond what's expected to deliver more than is needed, wanted, or desired is a final grace note to the future's untethered world. Delight is a complex sensation and highly personal, but an experience is more likely to be delightful if it's unexpected, intentional, non-obvious, not monetized, and easy to use.

PLEASANTLY SURPRISING

Delight is an emotional reaction to exceeded expectations. Having not lost my progress editing photographs was delightful partially because living through previous generations of software that didn't save my work (as yes, I know I should do) has taught me to expect the worse. Expecting an unpleasant experience and having a neutral one or expecting a neutral experience and having a good one exceeds our expectations. It goes "above and beyond." Functionality isn't enough to generate delight. Unless you've been struggling to fix a broken thing, having it operate as expected creates no surprise and, therefore, no delight.

CHAMBERLAIN GROUP

I expect a garage door opener to open my garage door. I don't expect it to close it for me if I forget. Chamberlain Group (which owns the garage opener brands Chamberlain, LiftMaster, Merlin, and Grifco, among others) has taken advantage of untethered technology to do just that, and it delights me.

They've even gone so far as to brand that particular piece of their products "myQ."

A garage door opener is a machine that, by definition, goes in my garage and stays there. It's primarily a piece of hardware, not a software product. But Chamberlain is unseating Genie, the long-time king of garage door openers, by challenging that idea. It has done this by making its openers Wi-Fi-enabled and capitalizing on that connectivity to deliver value and delight.

Chamberlain's door openers use a geofence (a GPS-enabled operation that creates a virtual perimeter in actual space) to monitor when my car enters or leaves a predefined area around my garage. They integrate powerfully with brands like Tesla to allow them to open the garage as I approach and close it behind me once I'm inside. This is handy but insufficient for delight. It's pretty much what you expect in a futuristic garage door opener.

What makes Chamberlain delightful is the things you don't expect. Most automatic garage doors will not close if they detect an object in the door's path. If I dropped my gym bag on my way out, Chamberlain, like Genie, won't close on it and crush my sneakers. Only Chamberlain will send me a text message through its app on my smartphone or smartwatch. I can then use that app to check and see what's keeping the door open (some models even come with a built-in camera!) and close it remotely if my wife picked up my gym bag for me.

Via its integration with home security systems like Ring and Vivint, I can also remotely open the door for the neighbor kid who comes by and walks my dog or generate (and later deactivate) a code she can use to open it herself. I can even automatically allow Amazon delivery drivers to leave packages safely inside my garage. They scan a code on their Amazon delivery app, and myQ confirms their location at my house and opens the door. Walmart's grocery delivery drivers can do the same thing.

Chamberlain's myQ technology even integrates with IFTTT, which allows it to integrate directly with dozens of other untethered products such as Philips Hue, Amazon Alexa, and even Midea air conditioners. Who wouldn't want their garage door to turn on their AC for them as they arrive home or turn it off to save money while they are at work? The Genie Company lost (and is still losing) market share because it didn't recognize what untethered technology made possible, while Chamberlain used it to create features that deliver unexpected value and, thus, delight.

INTENTIONALLY ALIGNED

Imagine it's mid-November, and you're starting to think about holiday gift-giving. Price and other things being equal, which would you rather give: a gift card, a selection of gourmet chocolates, or tickets to the person's favorite performer? Unless your friend is a certified chocoholic with an outstanding metabolism, my guess is that you'll give the tickets because they're the most thoughtful.

Most people like chocolate, but not everyone does (poor, misguided souls that they are). Even devoted chocolate lovers may be trying to cut down on their sugar intake. In giving chocolate, you run a risk of giving your friend something they'd really rather not have. Gift cards are a much safer bet since you know the recipient will end up getting themselves something they want. Still, even though they're always appreciated, they rarely feel personal.

What makes the tickets such a great gift is that they're both personal and practical—they're aligned with your friend's interests, needs, wants, and desires. While all three possible gifts speak to your intention to do something generous, only the tickets are delightful because they're well-aligned.

In the world of untethered design, this attribute of delight may be easier to spot in its absence. I mentioned Panera's incentive program in the previous chapter as a great example of meeting people where they are. I often stop there for my morning coffee, but I never eat lunch there. The occasional free coupon for a cookie or breakfast muffin is delightful. A free salad would not be. Not because I don't like salad but because it would feel like a transparent attempt to get me to buy something I don't want from Panera—lunch.

There may be no richer source of ham-fisted misalignment than the transparent failures of retargeting ads that assume if you're browsing for X, you might also be in the market for Y. A friend of mine was recently the surprised recipient of a free diaper in the mail. He's unmarried, childless, and content. It took him some serious thinking to trace the "free sample" back to a trip he'd made to a big-box store after he moved into a new but dirty office for the mega-pack of baby wipes he'd used to clean the place.

MAGICALLY DESIGNED

The art of stage magic is essentially that of taking something ordinary (like a deck of cards) and making it behave in a way the audience can't explain. Great design has the same effect—it makes the mundane mysterious. And the mysterious is both fascinating and delightful. It intrigues us and gives us pleasure. Thus, because humans are pleasure-seeking organisms, any truly well-designed thing automatically meets a desire.

Design, at its core, is problem-solving. What differentiates it from engineering is a shift of focus from task to experience—from "make the widget" to "make the widget meaningful." It's the difference between Dell and Apple. Good design requires empathy with the user's experience. Magical design elevates that experience into something a little bit mysterious.

With the best design, there's the immediate, positive surprise of exceeded expectations and the gratification of feeling aligned. Even once you've peeled back the layers to see the logical structure beneath the design, there's still something about it you can't quite explain. It's a bit like showing up on your friend's doorstep with those tickets after he's had a terrible day—exactly when he needs them. There's a *how did they know?* sense to it.

FRICTIONLESS TO ACCESS

If I needed to remove the battery from my Tesla and wheel it into the house to charge it, the experience would be significantly less delightful. When TiVo first came on the market, it didn't meet a single need, want, or desire of the TV-watching public that already owned a VCR. But it was delightfully easy to operate.

Features that reduce cognitive load or lower physical and learning barriers to use aren't inherently delightful. But features that would otherwise be pleasantly surprising, intentionally aligned, and magically designed but are difficult to use lose their ability to delight. You wouldn't ask your friend to walk miles in the rain to pick up his tickets at the box office four hours before the start of the show.

"FREE"

People don't mind paying more for an Apple than a Dell, but there's a limit to the price differential before they expect additional functionality. A free mint on your pillow is delightful. The bottle of water on the bedside table that shows up as a five-dollar room service fee is not.

Top-ranked hotels earn at least some of their five-star reputations by providing delight. Their staff goes "over and above." If you ask for restaurant recommendations, they offer to book a reservation for you. If you ask for walking directions, they give you a map. When they notice you've drunk all your in-room coffee, they restock with extras. And they don't do it for tips. They do things that add value to your experience without expecting a direct reward. This is not to say they don't benefit. They want you to come back and stay with them again. It's good for business if you tell a friend or tweet about your experience. And of course, the price of the room covers some of the additional goodies, but if you still leave feeling like you got more than you paid for, you'll be delighted.

UNCONSTRAINED DELIGHT

If you asked Genie about its business model, I bet it would say it built garage door openers. If you asked Chamberlain the same question, it would have a different answer. Both companies make garage door openers, but they think about the problem they solve differently. Genie opens garage doors. Chamberlain makes your home safer. Because Chamberlain frames its product as part of a larger picture, it's able to deliver delight. It's a bit like being asked to design a cheaper soup can when a box might be the lower cost, easier to stack, and more environmentally friendly answer to shelving minestrone. Defining the problem as "better soup can" rather than "better soup container" imposes constraints on the solution.

> Redefining the problem opens opportunities to provide delight.

This kind of out-of-the-box (out of the can?) thinking requires a different approach to the distribution of effort. To deliver delight, you can't just ask consumers what they want and provide it. There's nothing surprising about that. To create something delightful, companies need to shoulder the burden of discovery themselves. In the same way that your gift of tickets is more delightful if your friend didn't know their favorite act was coming to town, companies can deliver something people didn't know they wanted. If the company goes a step deeper, it can gain insights into the lives and goals of its consumers that the consumers aren't aware of. At the intersection of these deep consumer insights and the company's business and financial goals, opportunities to deliver delight come into focus.

Sometimes, this is as simple as asking why an additional time. To reinterpret the famous "faster horse" story, Ford understood that people hooked horses to carriages because they wanted to get somewhere more quickly than they could walk. If he'd focused on the horse rather than the speed, we might all still need saddles.

In a more mundane example, it's the difference between going to your marketing team and saying, "We need a new webpage, an email campaign, and a tracking link for the new product," and "We need to let people know about the new product." I call the first approach "solutioning" because a solution is bundled with the problem statement. It precludes other, possibly better and more interesting solutions.

On the opposite end of the spectrum from solutioning, there's the "It'd be cool if" approach to design. Here, the danger isn't that better or interesting solutions won't surface; it's that they won't solve anything.

In contrast, when Blizzard released the *Warlords of Draenor* expansion to its *World of Warcraft* game, it created an enormous replica of its iconic, highly stylized war hammer and put it through the roof of a yellow cab in Times Square. It looked like the aftermath of a truly epic battle, and people were fascinated. They posted pictures, wrote articles, and paid much more attention to it than they did to the movie posters that usually carry the weight of advertising a new film in Times Square. It was a compelling piece of guerilla marketing that wouldn't have happened if Blizzard had tasked an ad agency with print ads. It was cool, but if the hammer hadn't been recognizable (and, in fact, labeled), it would not have worked.

When looking for opportunities to provide delight, it's essential to introduce space between the desired outcome (tickets sold) and the standard method (posters) for delivering it. This is sometimes parsed as the difference between a strategic decision and a tactical one. It's tempting to leap straight to tactics. We like to take action and get things done, and tactical decisions lead to execution. But executing on an inferior strategy or on one based on unquestioned constraints isn't as efficient as it might initially seem, and it's rarely delightful.

Tesla, once again, provides an excellent example. Because it's a car company, it's concerned with the problem of cabin noise. Because it's a technology company, it solved that problem not by tasking its engineers with developing nearly soundproof glass or by creating more advanced door seals (as its classical car competitors did) but by creating a quieter cabin. How? It introduced an active noise cancellation system that identified the sound frequencies of road noise and generated a counter-frequency that eliminated it.

SUMMARY

The untethered consumer already has most of what they need, want, and desire. But delight, by definition, goes beyond that holy trinity. It sets a company apart from its competitors through a generosity of spirit that doesn't necessarily have a business case but is simply charming—proof of the company's untetheredness.

CONCLUSION

In the untethered future where everything is connected to everything, the traditional one-to-one relationship between goods or services provided to payment received is obsolete. To thrive in this new marketplace, companies need to avoid BlackBerry's mistake of seeing the future through the frame of the present.

When companies replace their outdated mobile-means-phone frame with an untethered one, they reconceptualize mobile as an environment in which they create and support non-transactional relationships with customers by understanding what they need, want, and desire and by providing value across the full range of platforms, devices, and apps. They meet people where they are with relevant messaging that's calibrated to the varying amounts of time, attention, and memory people have in different circumstances.

They recognize trust as the valuable currency it is and earn their customers' trust by being trustworthy. They also demonstrate their trust in their users by exerting less control, allowing them to personalize and customize their experience to their own tastes and needs. While the most successful such companies become the medium of people's mobile identities, every company can use untethered design to become more fully integrated with its customers' lives. And every business that adopts the seven principles of untethered design and the ethos of putting the customer's needs ahead of its own business goals can find opportunities to provide delight.

In *Untethered,* I've explored each of those seven principles individually and looked at the way they can intersect to deliver delight. Untethered design is:

- **Non-transactional:** Honoring the more personal, one-on-one relationship people have with their untethered devices by leveraging the smaller, more frequent, and more personal interactions makes it possible to foster human-to-human, rather than company-to-customer, relationships. While they're still business relationships, they differ from the older model by being more values-based and sustained over time.
- **Empathetic:** Recognizing the challenge of separating companies' goals from the goals of their customers, untethered companies strive for empathy. They begin their design processes by trying to understand their customers' needs, wants, and desires and only then look for innovative ways to serve their own goals by meeting those of their customers.
- **Application-, device- and platform-agnostic:** Keeping a company's focus on providing something its customers need, want, or desire puts a lower priority on getting credit for its work than on meeting users where they are. Rather than limiting the platforms it's on or driving traffic inorganically to its branded app or website, the company

prioritizes delivering value through whatever channels are most congenial to the people it's trying to serve.

- **Flexible:** By meeting people where they are, untethered companies recognize that *where* dictates *how* they interact and design accordingly. They leverage untethered technology's flexibility to adapt to the levels of attention, amounts of time, and modes of interaction that vary across a user's day and location.

- **Personalized:** An untethered company is too empathetic with its users' needs, wants, and desires and too invested in having human relationships with them to force irrelevant information on them. Instead, it personalizes its messaging, timing it appropriately and getting permission to send it. It's not trying to be heard at any cost. It's trying to say something people want to hear when and how they would most like to hear it.

- **Relevant:** Untethered companies tailor what they deliver to their individual users by leveraging the one-to-one relationship people have with their personal devices and by creating profiles to maintain that one-to-oneness on devices that are shared. They also enhance users' agency and control by allowing them to create further personalization to deliver more of what people want how they want it.

- **Integrated:** When an untethered company becomes an inseparable part of its users' mobile identities, it may be almost invisible to them. Paradoxically, this is a good thing, increasing the company's value to its users and the users' loyalty and use of the company's products and services.

- **Delightful:** While there are both traditional companies that deliver delight and successful untethered companies that aren't particularly delightful, you can always look for opportunities to deliver delight once you understand its components: be intentional about providing easy-to-access, unexpected value in a way that feels generous and sincere.

BlackBerry's failure of vision in the face of Apple's recon-ceptualization of what a phone could be is a great cautionary tale, but Google provides a great point of contrast. It was no more prepared than BlackBerry, but it was able to pivot and produce the first Android smartphone quickly and competi-tively because it changed its framing. In fewer than three years, Google upended its entire strategy—a shockingly fast pivot for anything hardware-heavy.

Although the iPhone is obviously still massively popular, most of the world now uses Android devices. This is only partially due to the price point. Apple's "walled garden" philosophy of tight control over what's allowed on their devices makes their phones less attractive to many than Android's open-source program-ming. Because it's free to manufacturers, almost all handsets now ship with Android's OS.

Today, business leaders face the same challenge to reframe and re-envision what mobile means. Reconceptualize how you think about, design for, and fund your mobile strategy. It's not too late to get ahead of the untethered future, contribute to the exploration and expansion of what it makes possible, and start exploiting its differences and potential. Creating more coherent and enjoyable—even delightful—experiences for customers is not the exclusive purview of tech companies led by eccentric geniuses. Every leader can have the vision to think about the mobile-first future from an untethered frame.

ACKNOWLEDGMENTS

Thanks to my mother, Linda, who has always supported me and been an amazing role model. I blame her for my love of coffee, my unending patience, and my big heart.

Thanks to my dad, Rich, who has always been there for me. I blame him for my eventual love of fitness, my appreciation of history, and my can-do attitude.

Thanks to my wife, Rachel. She still loves me after twenty years, is my favorite person to be around, and is super cool and supportive of all my ~~shenanigans~~ endeavors.

Thanks to my uncle, Bart, for ensuring I don't take myself too seriously and for hammering in my love of coffee. Bart was the first in the family to write a book, so you could say that some of this book and its occasionally comedic tone are his fault.

Thank you to my family: Aunt Susan; my cousins Jess, Jonathan, and Josh; and my great-grandmother Great. (Yes, that was really her name.)

Thanks to Jason Li, Justin Nguyen, Vince Francoeur, Sam Shank, Arron Goolsbey, and the college group—Richmond, Weigy, Lateef, Darius, Dan James, DJ, Chad, and Craig—and the entire Blizzard Mobile team.